Farm Blacksmithing
A Textbook and Problem Book For Agricultural Students and Farmers

by John F. Friese

with an introduction by Roger Chambers

Self Reliance Books

Get more historic titles on animal and stock breeding, gardening and old fashioned skills by visiting us at:

http://selfreliancebooks.blogspot.com/

Introduction

I am pleased to present yet another title on Homesteading and Farm Life.

This volume is entitled "Farm Blacksmithing" and was published in 1921.

The work is in the Public Domain and is re-printed here in accordance with Federal Laws.

As with all reprinted books of this age that are intended to perfectly reproduce the original edition, considerable pains and effort had to be undertaken to correct fading and sometimes outright damage to existing proofs of this title. At times, this task is quite monumental, requiring an almost total "rebuilding" of some pages from digital proofs of multiple copies. Despite this, imperfections still sometimes exist in the final proof and may detract from the visual appearance of the text.

I hope you enjoy reading this book as much as I enjoyed making it available to readers again.

Roger Chambers

Fig. 11—C. E. COLBURN'S FARM AND STOCK BARN

1

Fig. 19—MR. LAWSON VALENTINE'S BARN, "HOUGHTON FARM," MOUNTAINVILLE, N. Y.

PREFACE

THIS book is the direct outgrowth of the author's experiences in teaching farm blacksmithing to farm boys. The objects, of which the process of making is described in the book, all center about farm activities.

The purpose of the book is two-fold. It is intended as an aid to a teacher of farm blacksmithing, being a source of problems, and suggestive of how the work may be carried forward.

In addition to being a "What" and "How" book for instructors, the author constantly had in mind its direct use by farmers as a means of self-instruction in blacksmithing. This accounts for certain details and some repetitions. Because of its intended use as a book of self-instruction the author has taken little for granted.

Farm blacksmithing is a rougher type of work than that expected of a city smith on automobile or auto truck work. The important question is: "Does it fit and is it strong enough?" In welding the question is not, "How does it look?" but "Will it hold?"

The dimensions given on the drawings in this book are such that the work will be strong enough and of a size in general use. The sizes should, however, be modified when individual needs require. There is real value in making these new objects to dimensions, because only in that way will a person learn how to make duplicate parts for repairs.

Repair work in schools is only touched on briefly in this book. It should be stressed as much as possible, the students bringing broken parts of implements, etc., to school to repair. However, one who has forged most of the articles, as shown in this book, should be able to do almost any kind of blacksmith repair work on a farm.

JOHN F. FRIESE.

St. Cloud, Minn.
May 1, 1920.

ACKNOWLEDGMENT

TO THE farmer boys, my students in farm blacksmithing, at the Technical High School, St. Cloud, Minn., grateful acknowledgment is made. Without their help and inspiration this book would not have been possible.

J. F. F.

CONTENTS

Mild Steel—Wrought Iron
Tool Steel—Cast Iron

Hardening—Tempering—Hardening and Tempering the Cold Chisel—
Tools and Their Tempering Colors—Annealing—Case Hardening

PETER TUMBLEDOWN'S IMPLEMENT "GRAVEYARD." AVOID THIS BY HAVING A SHOP AND A KNOWLEDGE OF BLACKSMITHING.

TOOLS FOR THE FARM BLACKSMITH SHOP

IN EQUIPPING a farm blacksmith shop several things should be taken into consideration. The initial cost of the equipment is always an important item. The amount of work likely to be done in the shop, the character of the work, etc., should also be considered. Fig. 1 illustrates a group of tools which is considered a minimum equipment for starting a farm blacksmith shop. All simple forging of iron, and welding, can be done with these tools.

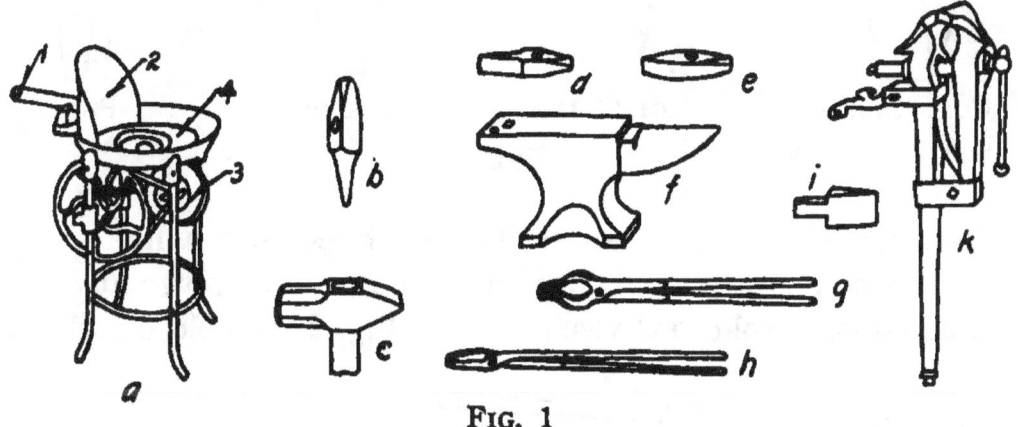

FIG. 1

The forge *a* has an 18 x 4-inch hearth. The fan is operated by a hand lever. Many larger and better hand forges are made, but this one will answer the purpose very well. Fans have now practically supplanted bellows for draft, giving as they do, a more steady draft. The punch *b* has a handle. A common size of punch is the one having a ⅜-inch point. The hand punch illustrated in Fig. 3 serves well for punching, especially for smaller sizes of holes. The hand hammer or blacksmith hammer is shown at *c*, Fig. 1. One weighing 2 pounds is a good size for general work. The machinist's or ball pein hammer of similar weight is now frequently preferred to the one shown in the illustration. The 1⅛-inch hot chisel, *d*, and 1⅛-inch cold chisel, *e*, are used respectively for cutting hot and cold iron. These are many times called hot and cold cutters. The 40-pound anvil, *f*, with tool steel face, will be found sufficient for most blacksmithing.   However, a full-sized blacksmith's anvil weighing about 120 pounds has considerable advantage over this one. A bolt tongs is illustrated at *g* and a straight lip tongs at *h*.

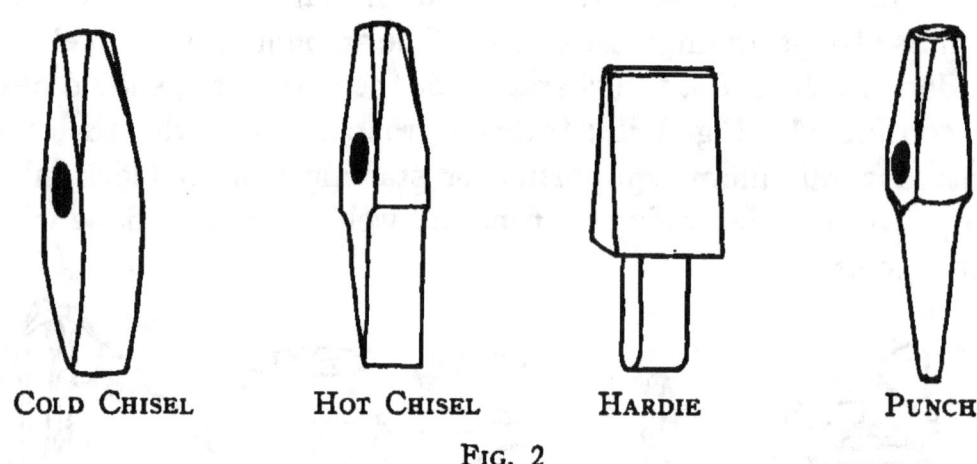

COLD CHISEL   HOT CHISEL   HARDIE   PUNCH

FIG. 2

The straight lip tongs can easily be formed into a link tongs, Fig. 3, when needed.   The hardie for the anvil is shown at *i*, and a blacksmith's solid box vise at *k*.   In Fig. 2 the cold chisel, hot

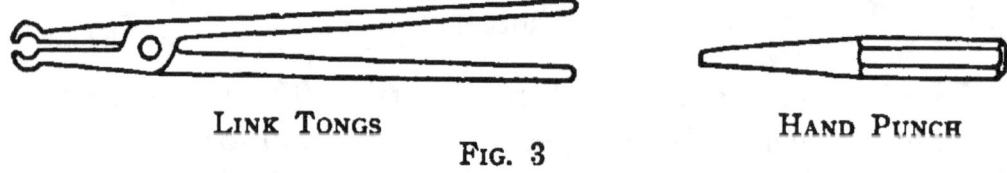

LINK TONGS   HAND PUNCH

FIG. 3

chisel, hardie and punch are again illustrated, so that the reader may gain a better idea of their shapes.   A rule, or better still a 12-inch steel square, is a necessary part of the equipment of a farm blacksmith shop.

In addition to the above tools the following ones will be found of considerable help, and time and labor saving.   The link tongs, Fig. 3, is used especially for holding round iron when it is being shaped and welded into circular articles such as links of chain and rings.   It is also used for holding iron when the ends are being squared or upset.   These terms will be made clear later.

A good link tongs is easily made from a straight-lip tongs by heating the lips or jaws, placing a piece of round iron between them near the end, and then hammering down the lips on both sides of the iron. A tongs known as the pick-up tongs, having much the same shape, can frequently be used in place of a link tongs.

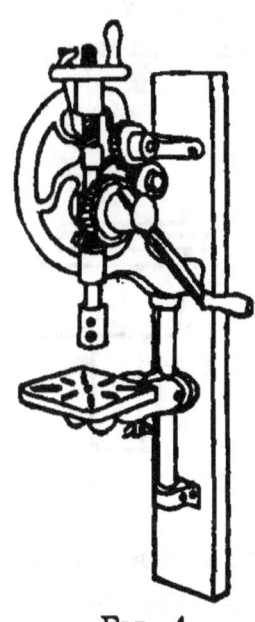

FIG. 4

A blacksmith's hand punch, Fig. 3, is a handy tool, and can readily be forged from a piece of octagonal tool steel. It is especially useful for smaller holes than can be punched with a punch with handle shown in Fig. 2. A center punch for marking dimensions on iron, and for centering before drilling with the post drill is a tool frequently needed. On Plate 27 will be found working drawings of a blacksmith's center punch and hand punch. Accompanying this plate are instructions for making them. A blacksmith's center punch with handle is manufactured.

The post drill, Fig. 4, is a great time-and-labor saver when many holes are to be made. Drilling is easier than punching, and the holes are better shaped. A drill is needed when the holes are to be tapped. The following drills are used most frequently: ¼ inch, ⅜ inch, ⁷⁄₁₆ inch, ½ inch, ⅝ inch, and ¾ inch.

Some second-cut files, a good monkey wrench or set of S wrenches, and a screw-plate set having taps and dies from ¼ inch to ¾ inch, are frequently needed. Where repair work is done on farm machinery, the cutting of threads on bolts and the tapping of holes is often necessary.

ADDITIONAL TOOLS USED IN SCHOOLS

FOR the farm blacksmith work in agricultural and technical schools of high school or college grade, tools in addition to those mentioned above are to be found. These tools are not absolutely necessary in the farm shop, but frequently better results can be obtained by using them, and a general knowledge of their use will do everyone some good.

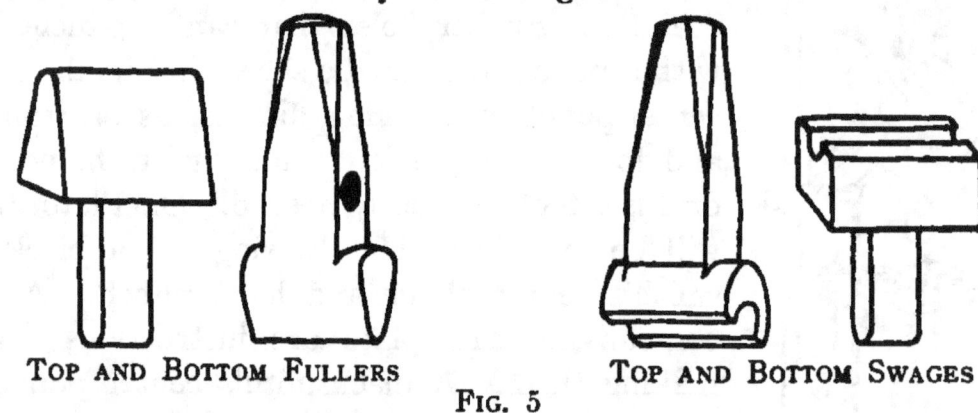

TOP AND BOTTOM FULLERS TOP AND BOTTOM SWAGES

FIG. 5

The top and bottom fullers, Fig. 5, are used to shape curved notches on one or both sides of a piece of iron. The bottom fuller fits into the tool hole in the anvil. The top fuller has a handle. Fullers are made in different sizes, the most common being ⅜ inch, ½ inch, ⅝ inch and ¾ inch.

Top and bottom swages, Fig. 5, are used for smoothing pieces of iron round after they have been forged approximately round. They are frequently used after two round pieces of iron have been welded. The diameters of the grooves are of various sizes, the most common being ⅜ inch, ½ inch, ⅝ inch and ¾ inch. A top fuller and a bottom swage of a larger size can be used to bend or curve sheet or flat iron.

The set hammer, Fig. 6, is used for shaping square corners. Set hammers range in size from 1¼ inches to 2 inches by eighths. The 1¼-inch size is common.

The flatter, Fig. 6, has a handle like the set hammer. It is used, as its name indicates to flatten flat surfaces after they have been shaped with the hand hammer.   Flatters range in size from 2 inches to 4 inches by quarters.

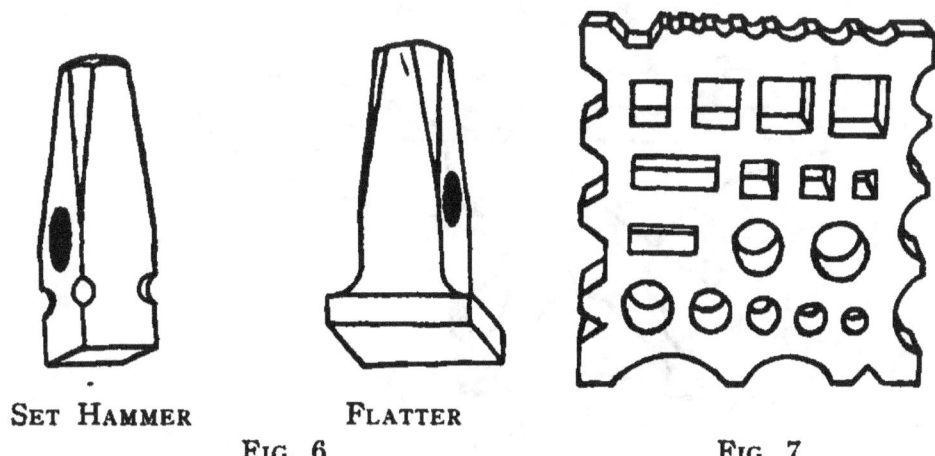

SET HAMMER   FLATTER

FIG. 6   FIG. 7

Cast iron swage blocks, Fig. 7, come in various sizes and shapes.   They are usually about 4 inches thick.   The block can be used in either an upright position or flat.   The various grooves, and round, square and oblong holes are used for shaping.   The holes can be used for punching, and for shouldering thick stock; also for hollowing sheet-metal.

In Fig. 8 is illustrated a blacksmith's mandrel or cone.   These are used for shaping rings and circular objects of all kinds when a true circle or part of a circle is desired.   Mandrels range in height from 32 inches to 50 inches and have top diameters of 1 inch, 2 inches and 5 inches.   A mandrel known as "Cheney's" has a slot running up the entire height. The slot is wide enough to admit the jaws of a tongs, and is advantageous when rings are being shaped.

FIG. 8

A person using a forge and anvil should know the names of the principal parts so that he can speak intelligently of them. The forge illustrated in Fig. 1 is known as a portable forge, and is the least expensive type. This forge is designed principally for farmers whose repair work is only occasional, and of a slight nature. The draft is furnished by the fan *3*. This forge is called

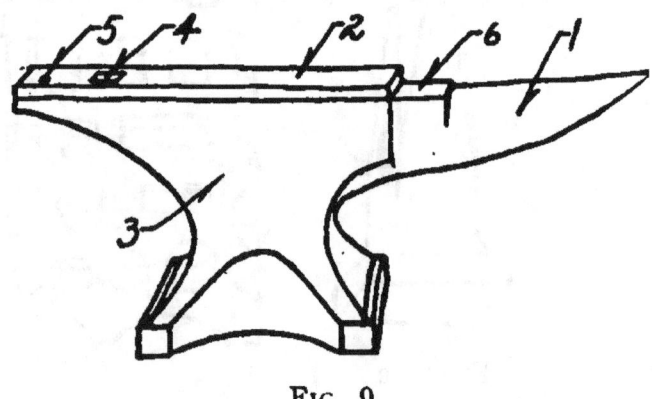

Fig. 9

a hand lever forge, because it is operated by a hand lever *1*. The shield is shown at *2*, and the hearth is at *4*. Where a considerable amount of rather large work is to be done a larger forge should be purchased. One that has a crank, geared direct to the fan at the rear, a half-hood, and a tuyere iron that produces a good draft, should be purchased. Several forges of this type are designed and manufactured primarily for farm use. Every forge should have four fire tools: a coal shovel, rake, dipper and straight poker.

The various parts of the anvil, Fig. 9, are named as follows: *1*, horn; *2*, face, usually of tool steel; *3*, body; *4*, tool hole; *5*, round or pritchel hole; *6*, base of horn.

BUILDING A FIRE

THE first thing one who is going to do blacksmithing must learn is to build a fire properly. Place a handful of shavings or paper in the firepot. Light and give a little draft and throw some fine blacksmith coal over it. Let this burn well and then push it to the center and add green coal all around it. Green coal is nothing more than blacksmith coal thoroly wetted. As this coal changes to coke push it to the center and add more green coal around the edge of the firepot, until you have a fire that is much like that illustrated in Fig. 10. The iron that

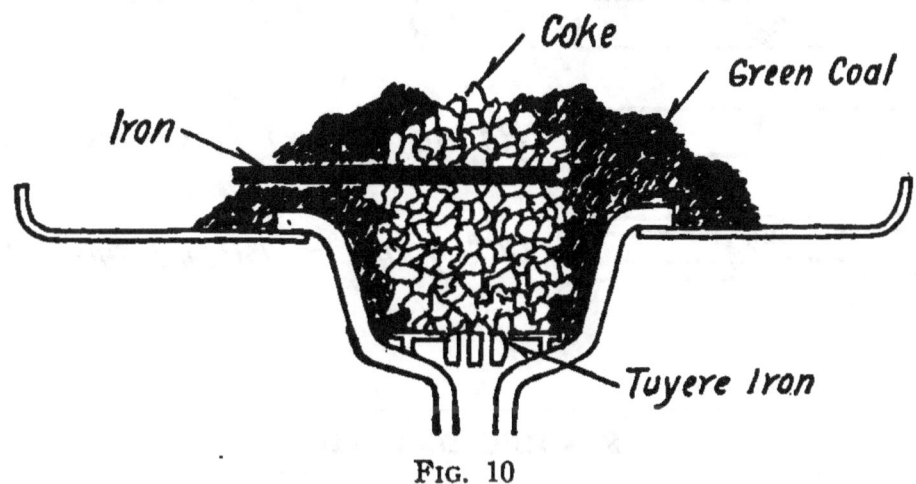

FIG. 10

is to be forged should be heated in burning *coke*, not coal. It should be placed in the fire *horizontal* as shown, and not toward the bottom of the firepot. The iron should have burning coke *below*, *around* and *above* it while heating.

A second fire is easier to build because some coke will always be left over. All that is required is to start this coke burning and heat the iron. The green coal is placed around the edges of the firepot so that the gas may be burned out of it and new coke continually be formed to take the place of that burned away.

A good blacksmith coal should be used. Tho slightly more expensive than ordinary soft coal it is cheaper in the end, because of time and labor saved. Blacksmith coal has very little or no sulphur in it. While iron can be heated with ordinary soft coal and shaped, it is out of the question to try to do welding with a coal that has sulphur in it.

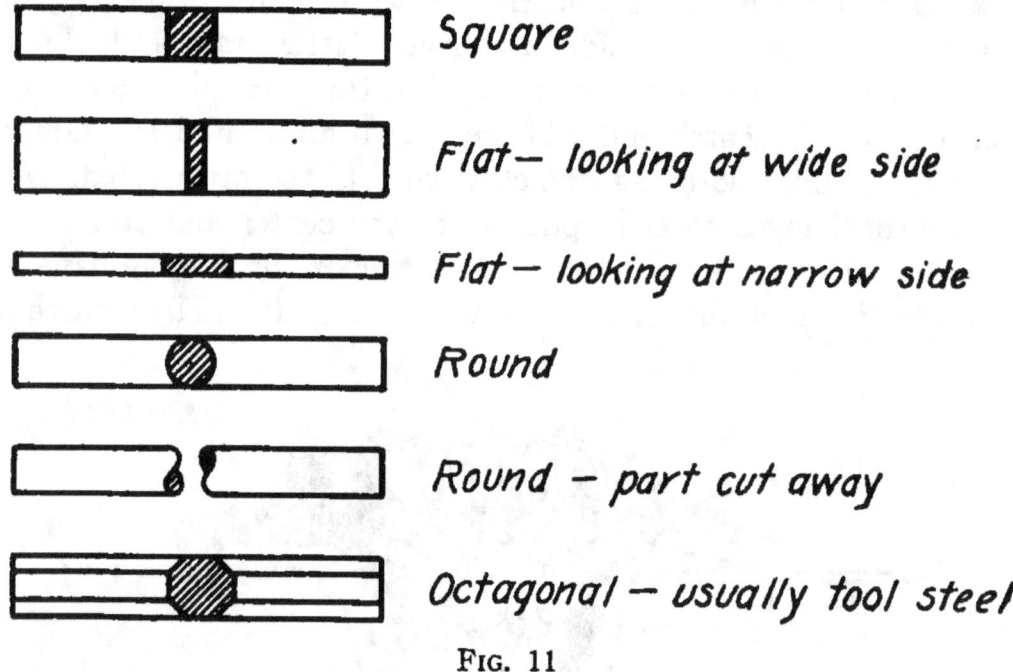

Square

Flat — looking at wide side

Flat — looking at narrow side

Round

Round — part cut away

Octagonal — usually tool steel

Fig. 11

READING DRAWINGS

FOR the reader who has had no experience in reading working drawings or blueprints the following brief explanation is offered · The drawings in this book are usually composed of one or more views of the object in its natural position. Sometimes the front view is given, sometimes the front and top, or front and side views. In other words, they are the views that one sees when looking at the object from the front, top or side of it.

In order to show the shape of the iron at certain places, cross-sections have been drawn on the object where the iron is of the shape indicated by the sectional part. If the iron were cut in two at the sectional point it would have the shape shown in the section. Fig. 11 may make this explanation a trifle clearer.

Important

BLACKSMITHING OPERATIONS

THE following operations are explained and illustrated in the places indicated. They have not been repeated in detail for each object in the book. They may only be named in the proper order under "How to Make." For the person who does not start at the beginning and make each object in the book, one after the other, it may be necessary to refer to the places where these operations are described and illustrated.

Squaring. See the Ring, Plate 1, and accompanying directions.

Shaping Points. See the Staples, Plate 2, and accompanying directions.

Bending. See Plates 1, 3 and 8 and accompanying directions.

Upsetting. See the Square Grab Hook, Plate 8, and directions.

Blackening. See the Ring, Plate 1, and directions.

Punching. See the Gate Hinge, Plate 7, and directions.

Drilling. See the Small Clevis, Plate 10, and directions.

Threading. See Plates 11 and 15, and accompanying directions.

Tapping. See Plates 7 and 11, and accompanying directions.

Heading. See the Welded Clevis, Plate 22, and Bolt Head, Plate 26.

Welding. See Plates 18, 19 and 20, and accompanying directions.

Hardening. See Plate 27, and accompanying directions, and page 87.

Tempering. See Plate 27, and accompanying directions, and page 87.

Case Hardening. See the Ice Tongs, Plate 28, and page 89.

RING

This ring is of size frequently used on a farm. By following the method given in the note on page 18 the amount of stock necessary for any ring may be found. Be sure to read what is written about tools in the first section of this book, unless you already know the names and uses of common blacksmith tools. Read all the directions before you begin work

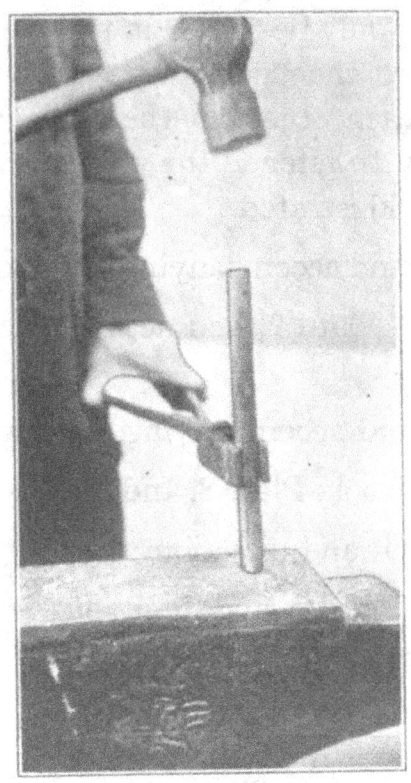 

FIG. 12. SQUARING AN END, ALSO UPSETTING.   FIG. 13. BENDING OVER THE HORN OF THE ANVIL.

How to Make the Ring

1. Cut a piece of stock to the dimensions given on Plate 1.
2. Heat to a cherry red, no hotter, and square both ends as shown in Fig. 12. Use a link tongs. In all blacksmithing always square the ends of your stock first.
3. Heat and bend one-third of the rod over the horn of the anvil as shown in Fig. 13.
4. Bend the other end in the same way.

RING

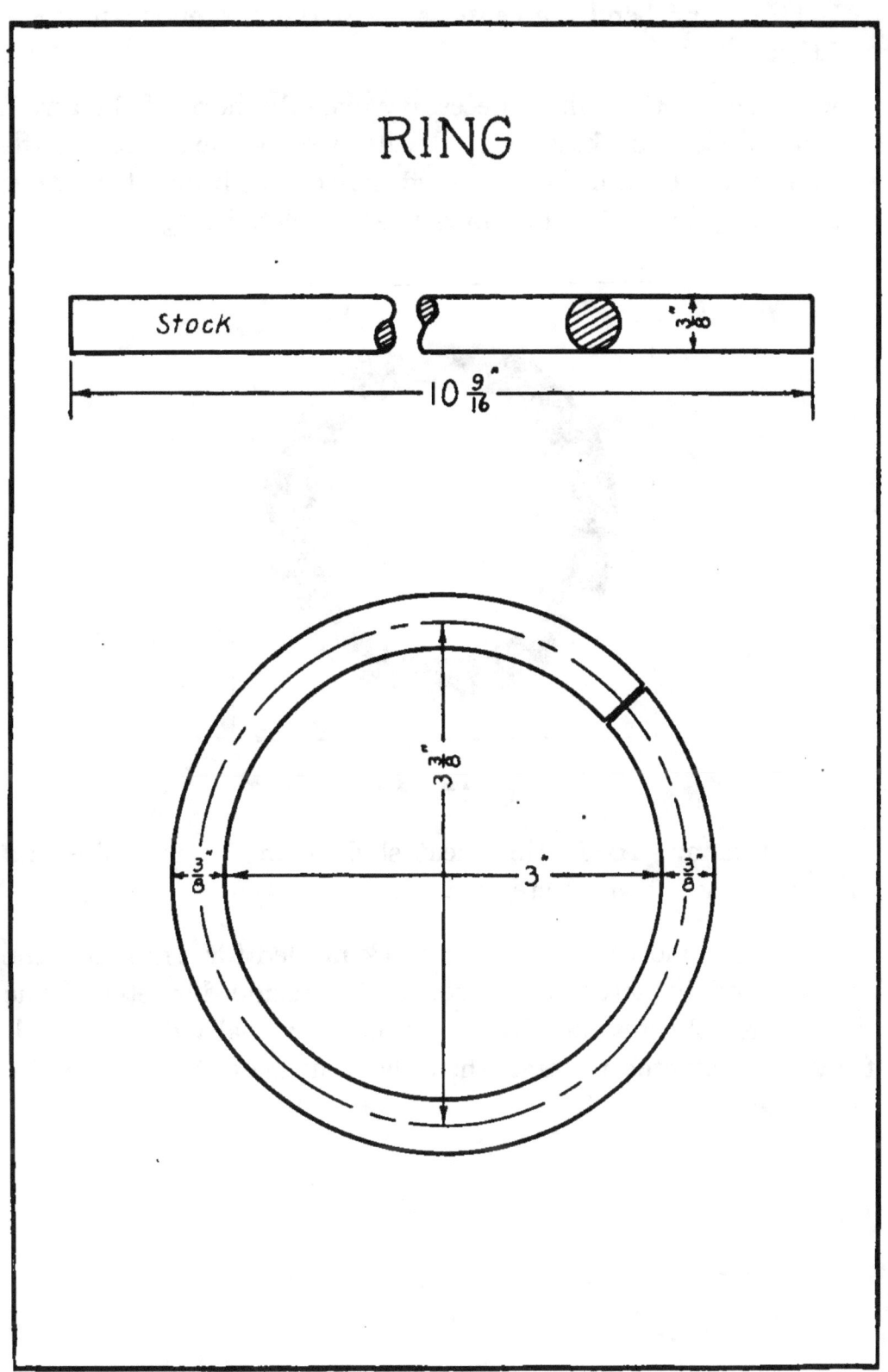

PLATE 1

5. Heat and bend the center so that the two ends will come together.

6. True the ring where necessary over the horn of the anvil, or a mandrel if you have one. If the two ends are filed a trifle just before the bending is completed they can be brought together in a better joint. Fig. 14 shows the completed ring.

FIG. 14. RING.

7. Blacken. To do this, heat slightly in a smoky fire and rub with oily cotton waste or cloth.

Note.—To find the amount of stock needed for a ring add the thickness or diameter of the iron to the inside diameter of the ring. Multiply this sum by $3\frac{1}{7}$. What you really do is to multiply the diameter running thru the center of the iron by $3\frac{1}{7}$. See Plate 1.

FLAT POINT STAPLE

Two staples such as this one, Plate 2, or the square pointed one, and the gate hook shown on Plate 3, will always keep a gate closed. Try to figure the amount of stock required. You have half of a ring, two straight parts 1 inch long and two flat points. Read this entire page before you begin work.

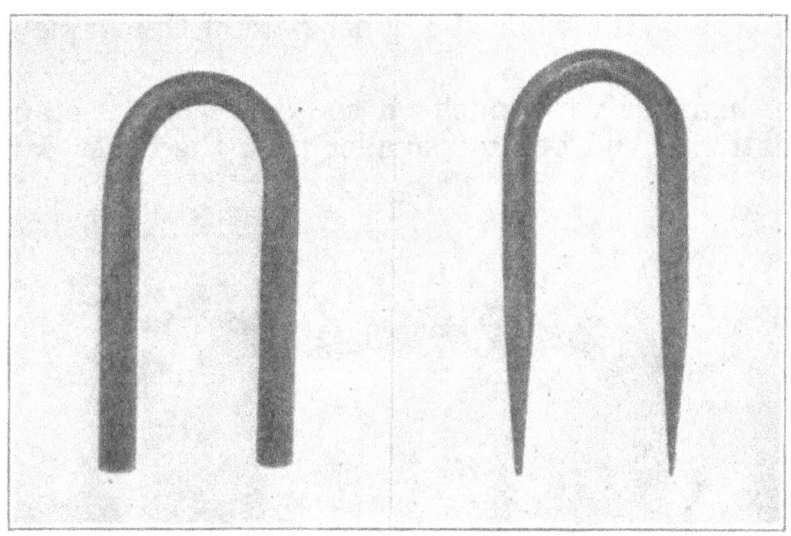

FIG. 15. SQUARE AND FLAT POINT STAPLES.

How to Make the Flat Point Staple

1. Square both ends.

2. Heat and hammer, "draw," both ends to a chisel edge working back *from* the points until they are just 1½ inches long. Hammer on both sides of the prongs, and always keep the under side flat on top of the anvil. Be sure to keep the points ¼ inch wide at the same time you are flattening them.

3. Bend at the center over the horn of the anvil as you did with the ring. Be sure to bend the points so that they take the position shown in Fig. 15 and Plate 2. Remember the two sides are just 1 inch apart all the way up.

4. Blacken.

SQUARE POINT STAPLE

1. Square both ends.

2. "Draw" both ends to square points, drawing the extreme points first and working back to the required length.

3. Bend at the center over the horn of the anvil.

4. Check up on your dimensions with the drawing and blacken. Fig. 15 shows the completed staple.

Notes.—As small iron heats quicker than large be careful not to "burn" the points while the upper part of the staple is being heated.

Do not hammer very much on the curve of the staple or it will be flattened and become smaller than the other parts

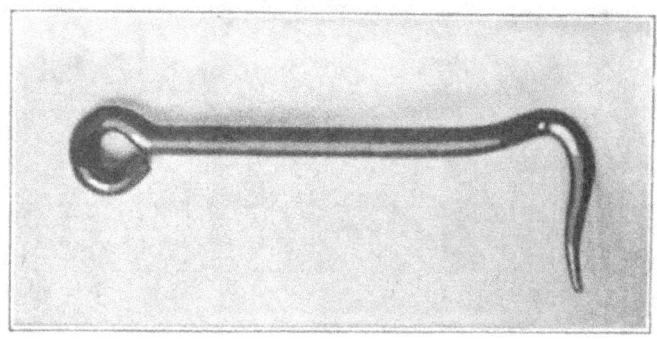

FIG. 16.   GATE HOOK.

GATE HOOK

This gate hook, Plate 3, is designed to be used with either of the staples shown on Plate 2.

How to Make the Gate Hook

1. Square ends and mark dimensions with center punch or chisel.

2. Draw a square point (as on the square point staple) 1¼ inches long working from the end of the stock backward. Hold with bolt tongs.

3. Round the point.

4. Bend the eye. Heat and clamp in the vise and hammer over the stock in the same manner as is shown for the hook in Fig. 17. Then bend over the horn of the anvil until the eye is shaped, as in Fig. 18.

STAPLES

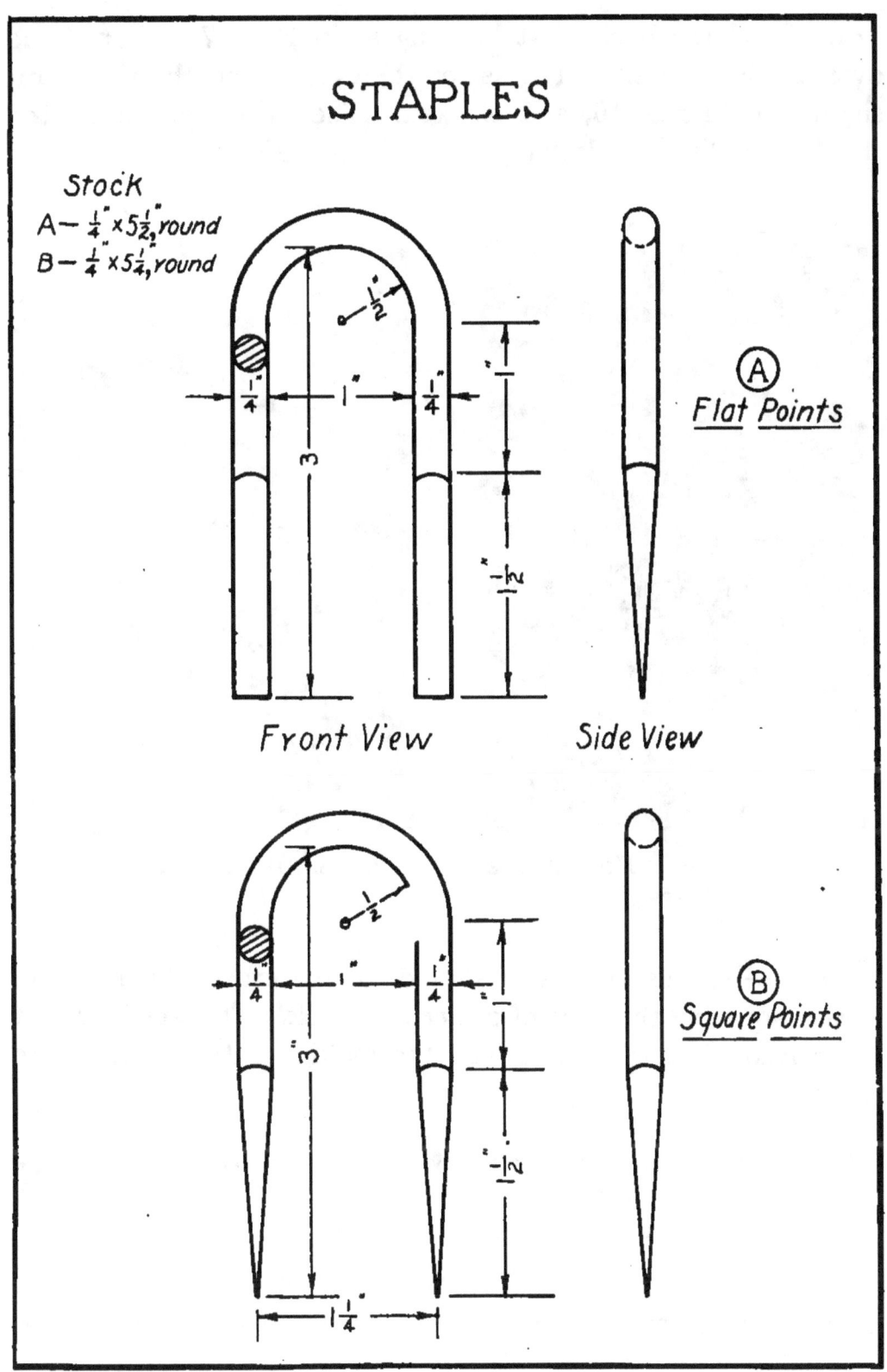

Stock
A — ¼" × 5½", round
B — ¼" × 5¼", round

Front View   Side View

Ⓐ
Flat Points

Ⓑ
Square Points

PLATE 2

5.   Bend the hook, first bending as in Fig. 17.   Then bend over the horn of the anvil as for the eye.   Bend the point out slightly as in Fig. 16, and blacken.   The gate hook completed should look like the drawing, Plate 3, and Fig. 18.

FIG. 17.   BENDING HOOK IN THE VISE.

FIG. 18.   BENDING THE POINT.

Notes.—By this time you should have formed the habit of always squaring the ends of a piece of stock.   This step will not be mentioned in the directions for making future articles, in this book.

The center of the eye should be in line with the center of the long body of the hook.

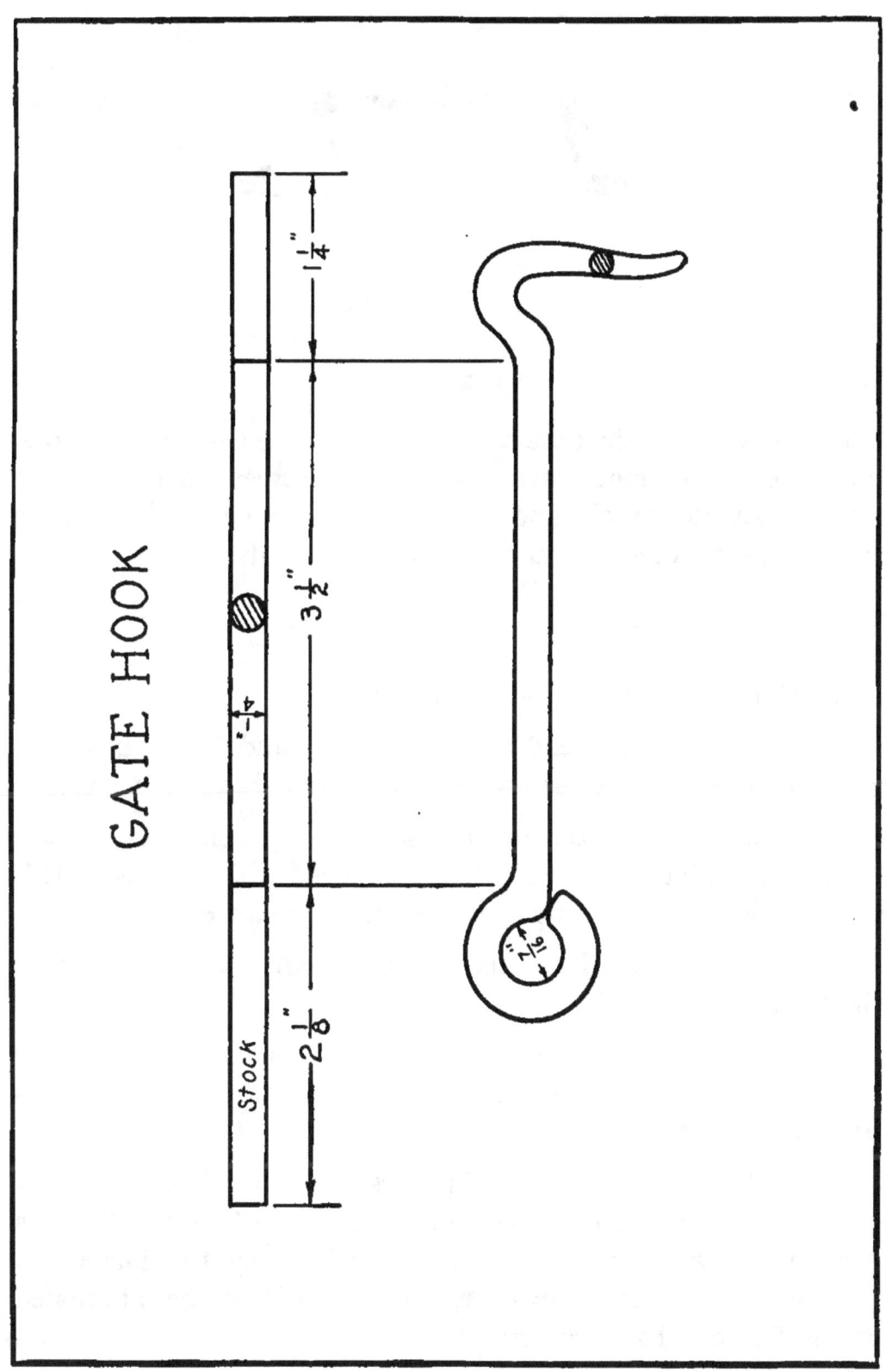

PLATE 3

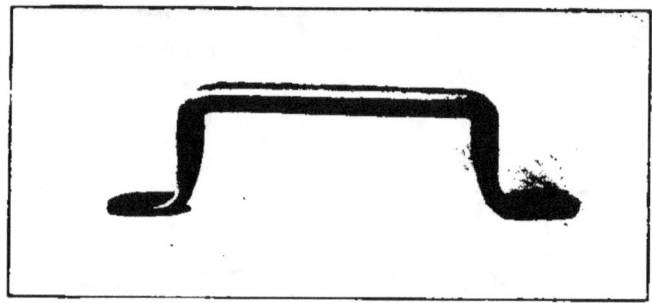

FIG. 19. DOOR PULL.

DOOR PULL

The door pull of the dimensions shown in the drawing, Plate 4, is suitable for a single barn door. For a screen door it can be made of smaller stock, and for a large sliding door it may be made large enough to accommodate two hands.

How to Make the Door Pull

1. Cut the stock to the dimensions given on Plate 4.

2. Flatten out each end to $\frac{3}{16}$ inch thick and $1\frac{1}{4}$ inches long. Shape as indicated in the drawing. Hold with a bolt tongs.

3. Mark with center-punch and drill $\frac{1}{4}$-inch holes in the ends as indicated. These might be punched with a blacksmith's punch. See "How to Punch Holes" under Gate Hinge, Plate 7.

4. Mark and bend the handle as at AA. over the horn of the anvil.

5. Mark and bend again at BB. In Fig. 19 is shown the completed door pull, the one in the illustration, however, was bent in a vise at AA.

Note.—The oily waste used in blackening—see 7 under "How to Make a Ring"—covers the iron with a thin film of oil. This helps to prevent rusting. It is a good policy to blacken all forgings in this way. However, this step will not be mentioned in the forging of future articles.

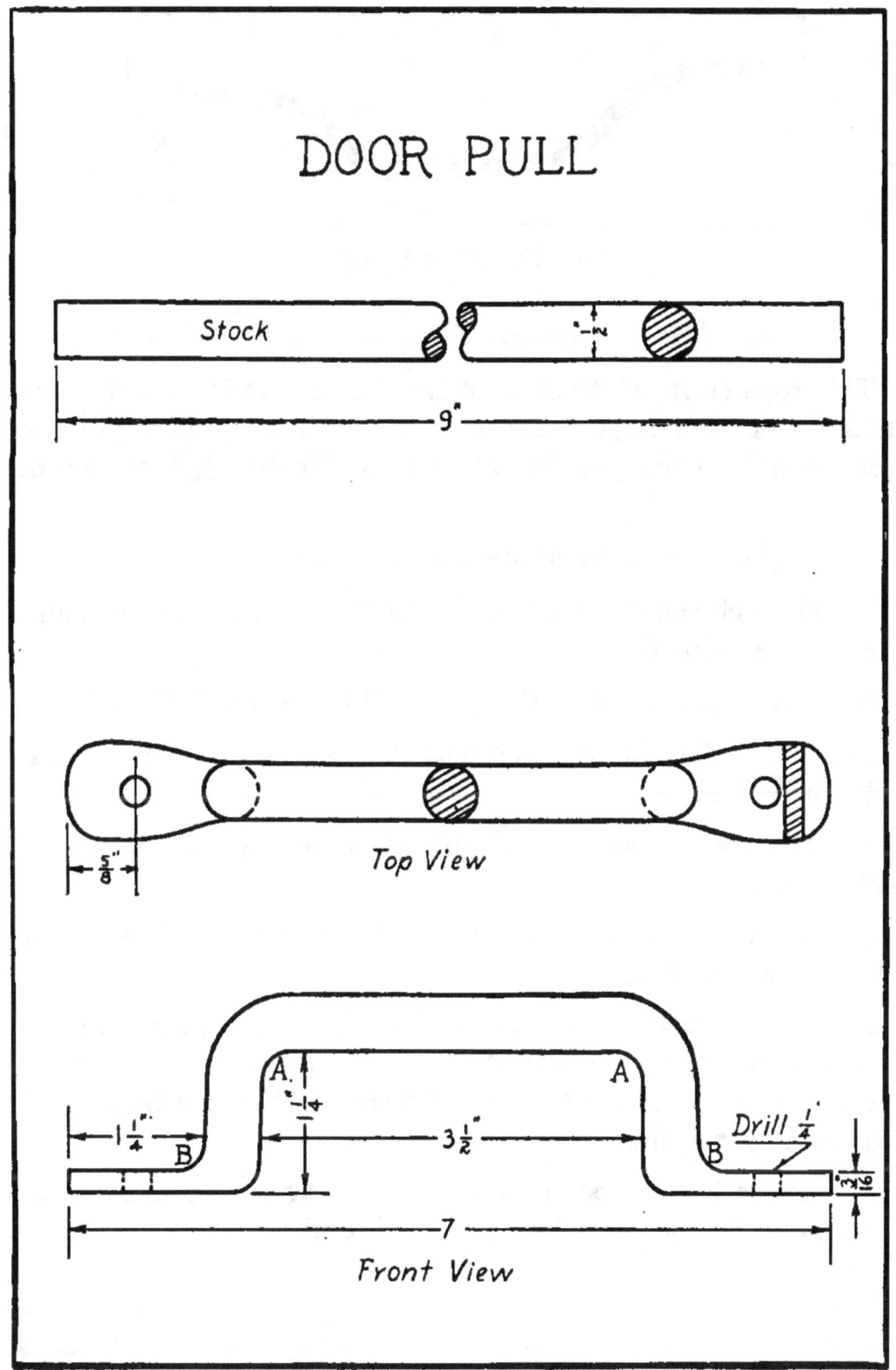

PLATE 4

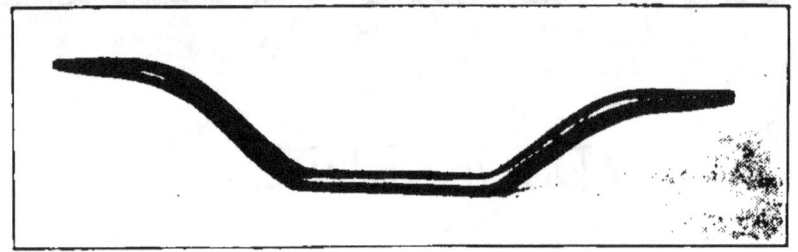

FIG. 20. ROPE CLEAT.

ROPE CLEAT

This rope cleat, Plate 5, is of small size. It is intended for small ropes. For large ropes or where there is great strain the stock should be correspondingly larger, as ⅝ inch or ¾ inch round.

How to Make the Rope Cleat

1.   On each end draw a blunted square point 2½ inches long. See Square Point Staple.

2.   Round the points making them 3 inches long, Plate 5.

3.   Flatten 3 inches of the rod at the center as shown, making it ³⁄₁₆ inch thick.

4.   Mark with center-punch and drill two ¼-inch holes 1½ inches apart.

5.   Bend the two ends or horns to the shape indicated on Plate 5 and in Fig. 20.

Notes.—If the link, bolt or straight-lip tongs are not of the correct size for any particular job, heat and shape them on a piece of iron to make them the right size. They can be shaped over and over again.

When heating a piece of iron, turn it over occasionally, so that the entire piece will be heated uniformly.

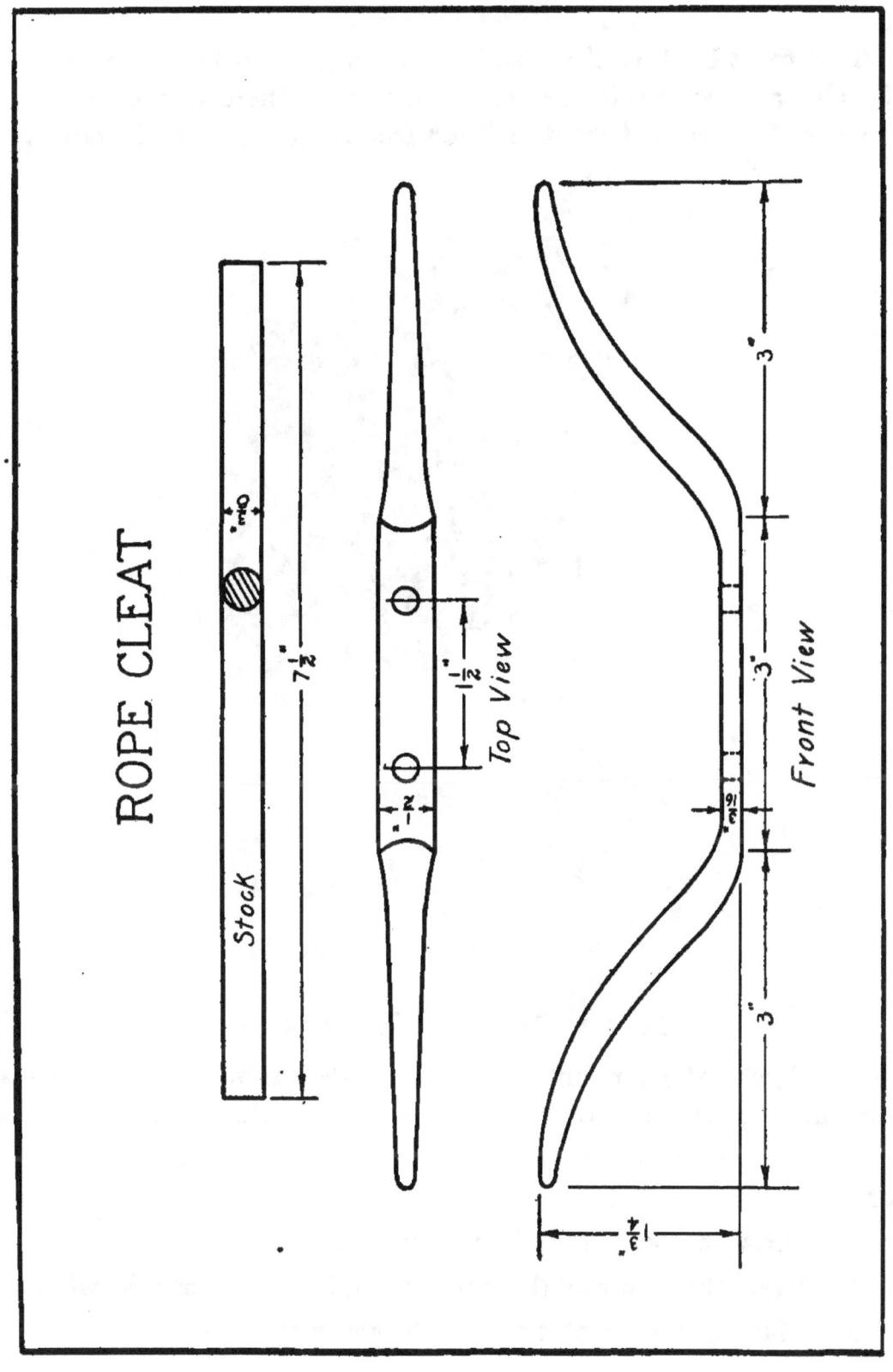

ROPE CLEAT

Stock

$7\frac{1}{2}''$

$\frac{3}{8}''$

$1\frac{1}{2}''$

$\frac{2''}{}$

$1''$

Top View

3″

3″

3″

$\frac{3}{16}''$

$1\frac{3}{4}''$

Front View

PLATE 5

MEAT HOOK

This meat hook is of a size that fits over a surfaced 2-inch x 4-inch or 2-inch x 6-inch piece of plank. Where a thinner or thicker plank is used the dimension from C to D, Plate 6, should be changed accordingly.

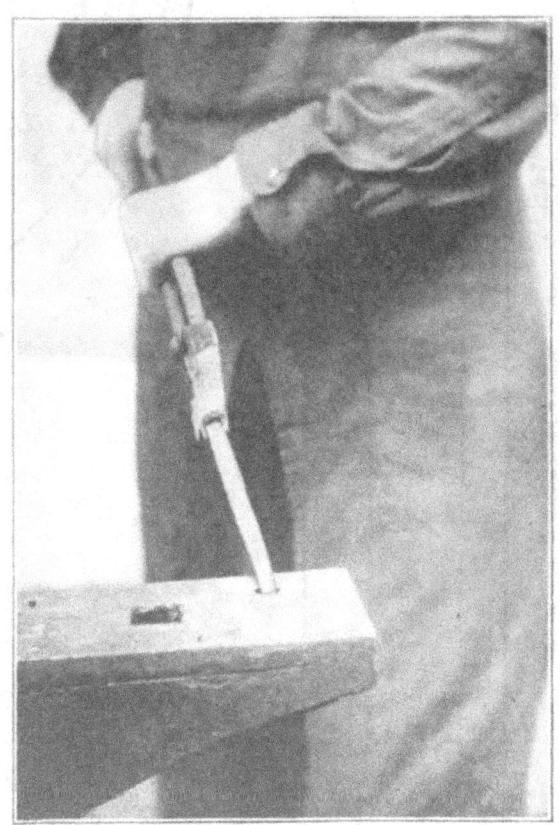

FIG. 21. BENDING IN THE PRITCHEL HOLE

How to Make a Meat Hook

1. Figure the amount of stock needed from the working drawing. Remember that corners require stock, also.

2. Draw a square point 1½ inches long on one end, just as on the square point staple.

3. Draw the point round to 2 inches long.

4. Draw the iron round from E to B, 3 inches; see drawing

5. Mark A with center punch, 3 inches from point.

MEAT HOOK

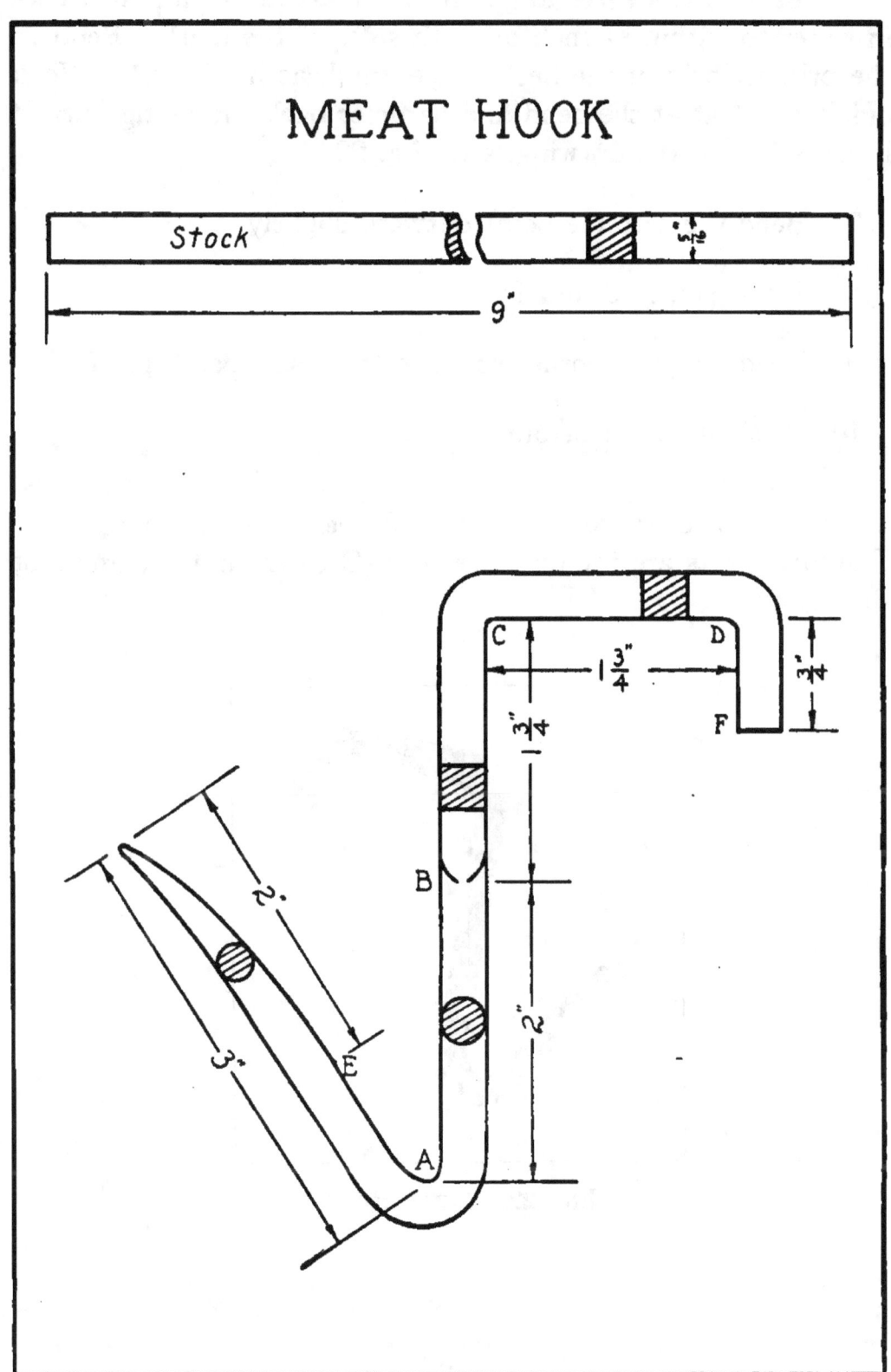

PLATE 6

6. Bend a 45-degree angle at A. Heat at A and then cool in water to within ¾ inch on both sides of the mark. Bend in the pritchel hole in the heel of the anvil, as in Fig. 21. Work quickly! Heat at the bend and hammer back the prong until it is shaped as in the drawing, and Fig. 22.

7. Bend the extreme point outward slightly.

8. Mark points C and D.

9. Bend a square corner at C, in the vise. See Fig. 17.

10. Bend at D, as before.

Note.—The distance from D to F may vary from ¾ inch. The vital parts are the distance from C to D, and the angle of the prong.

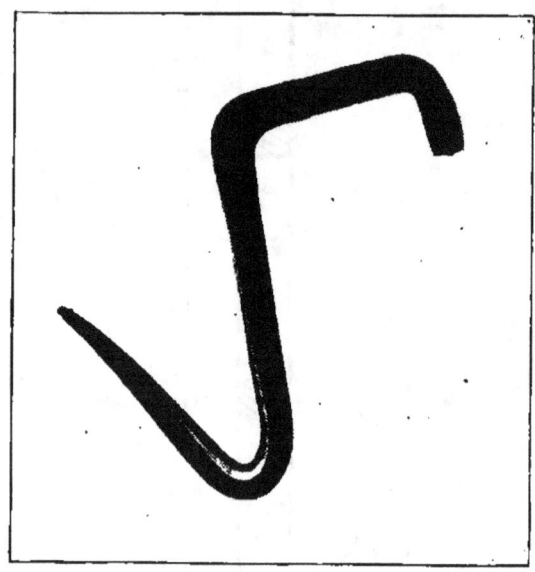

FIG. 22. MEAT HOOK.

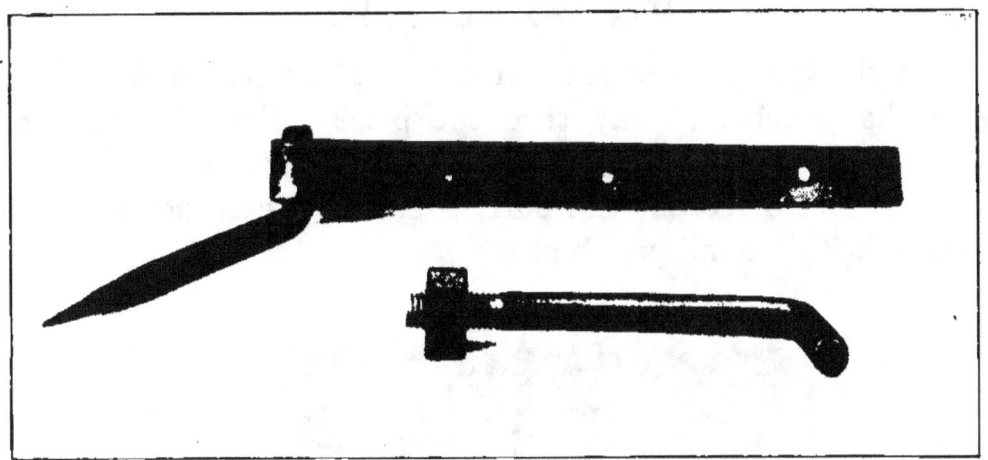

FIG. 23. GATE HINGE.

GATE HINGE

Two new forging operations, punching and riveting, are used in making the gate hinge. The stock is listed on the drawing, Plate 7.

How to Make the Gate Hinge

1. On the hook part of the hinge, A, draw a square point 2½ inches long.

2. Bend in the vise or over the anvil to form the 1½-inch hook. Use a set hammer, if you have one, to shape the up-turned part perfectly round at the place it is bent. If this part is not round the hinge will "stick."

3. On the strap part, B, mark off the centers for the holes and the amount of stock used to go around A.

4. Heat and bend the strap about A making a loose fit so that it will turn easily.

5. If the holes are to be drilled this should be done now. Three ¼-inch holes for attaching the hinge to the gate, and a ¼-inch hole for the rivet, are needed.

6. Heat a rivet of the proper size (make one if necessary), and rivet together. Fig. 23 shows the completed hinge, with the holes punched instead of drilled.

How to Punch Holes

Heat the iron to a bright red heat. Place on the anvil and drive the punch half-way thru the piece. Turn it over and punch half-way thru again. Then place over pritchel hole or tool hole and drive out the burr, Fig. 24. These holes can be punched with a hand punch also.

FIG. 24. PUNCHING—DRIVING OUT THE BURR
OVER THE PRITCHEL HOLE.

Notes.—The hook part, A, may be made with the end threaded instead of pointed. See drawing and Fig. 23. In Fig. 23 is shown a nut which was hand made. A piece of square stock was upset to the proper size, a ½-inch hole drilled and the hole tapped with a ⅝-inch tap.

Be sure the end is perfectly round before you try to thread it.

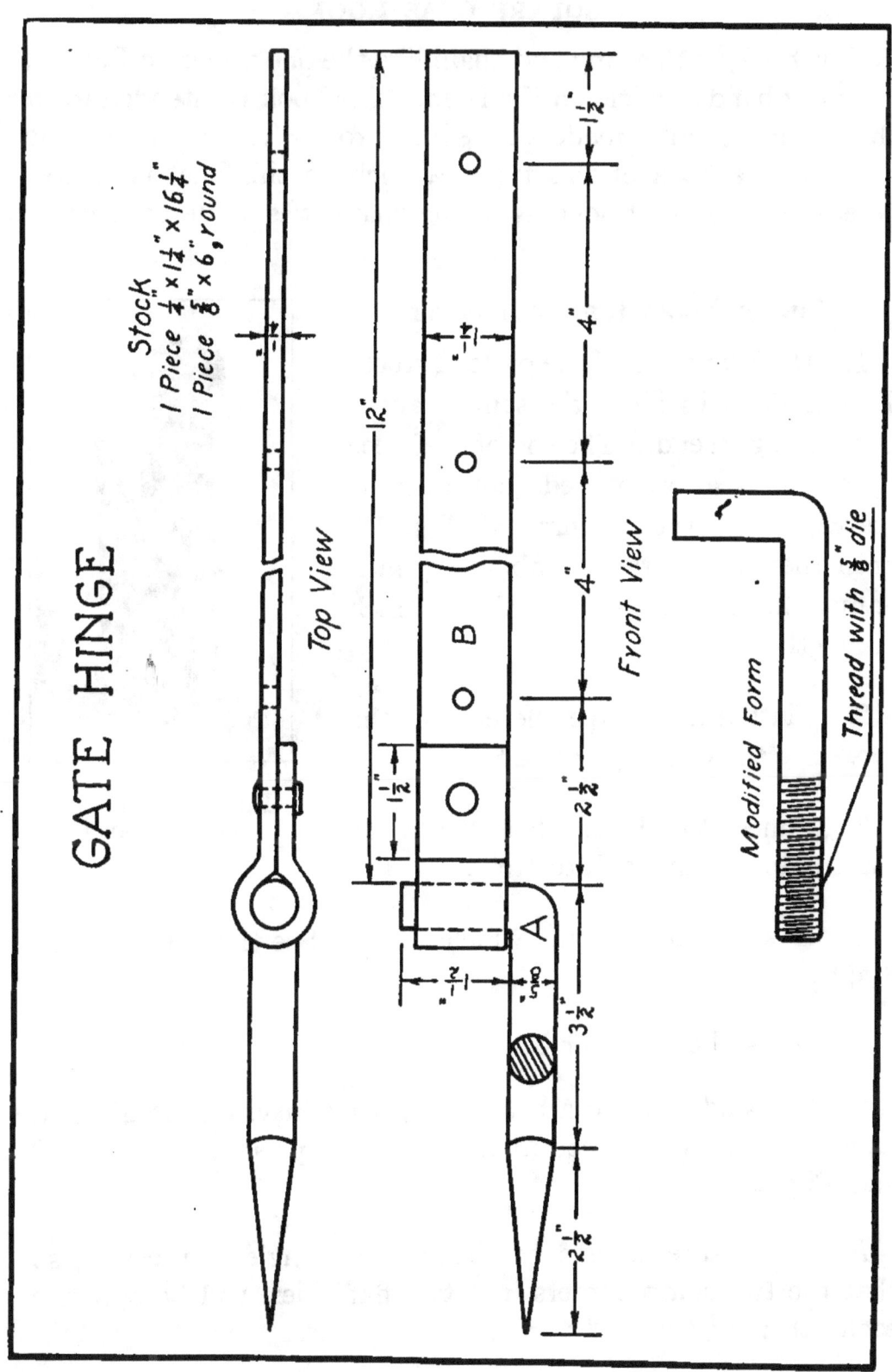

GATE HINGE

Stock
1 Piece ¼"×1¼"×16¼"
1 Piece ⅝"×6, round

Top View

Front View

12"

1½"

4"

4"

4"

2½"

3½"

2½"

1½"

B

A

Modified Form

Thread with ⅝" die

PLATE 7

SQUARE GRAB HOOK

This hook, with a ring and chain like the one shown in Fig. 41, is a very handy article on the farm. The hook is intended for a chain having links made of ⅜-inch iron. The width of the opening in a hook of this type (⁷⁄₁₆ inch on this one) is always determined by the thickness of the chain it is to be used with.

How to Make the Grab Hook

1. Upset one end of the piece of stock. This is done in much the same manner as squaring an end. The end of the piece is heated to a bright red (not until it sparks). It is then hammered, Fig. 12, until the end enlarges to about ¾ inch or more in thickness and for about 1 inch in length.

2. Flatten and shape the end for the eye. See the drawing, Plate 8.

3. Punch the hole. See "How to Punch Holes" under Gate Hinge.

FIG. 25. SQUARE GRAB HOOK.

4. Draw a square point 2 inches long. See the Square Point Staple.

5. Mark the place for bending.

6. To bend, heat to a bright red, place in vise at mark, and hammer over with a wood mallet to the shape shown on Plate 8 and Fig. 25.

Note.—Be sure to bend the hook at an angle in the vise, so that the two sharp corners, not two flat sides, will be opposite each other.

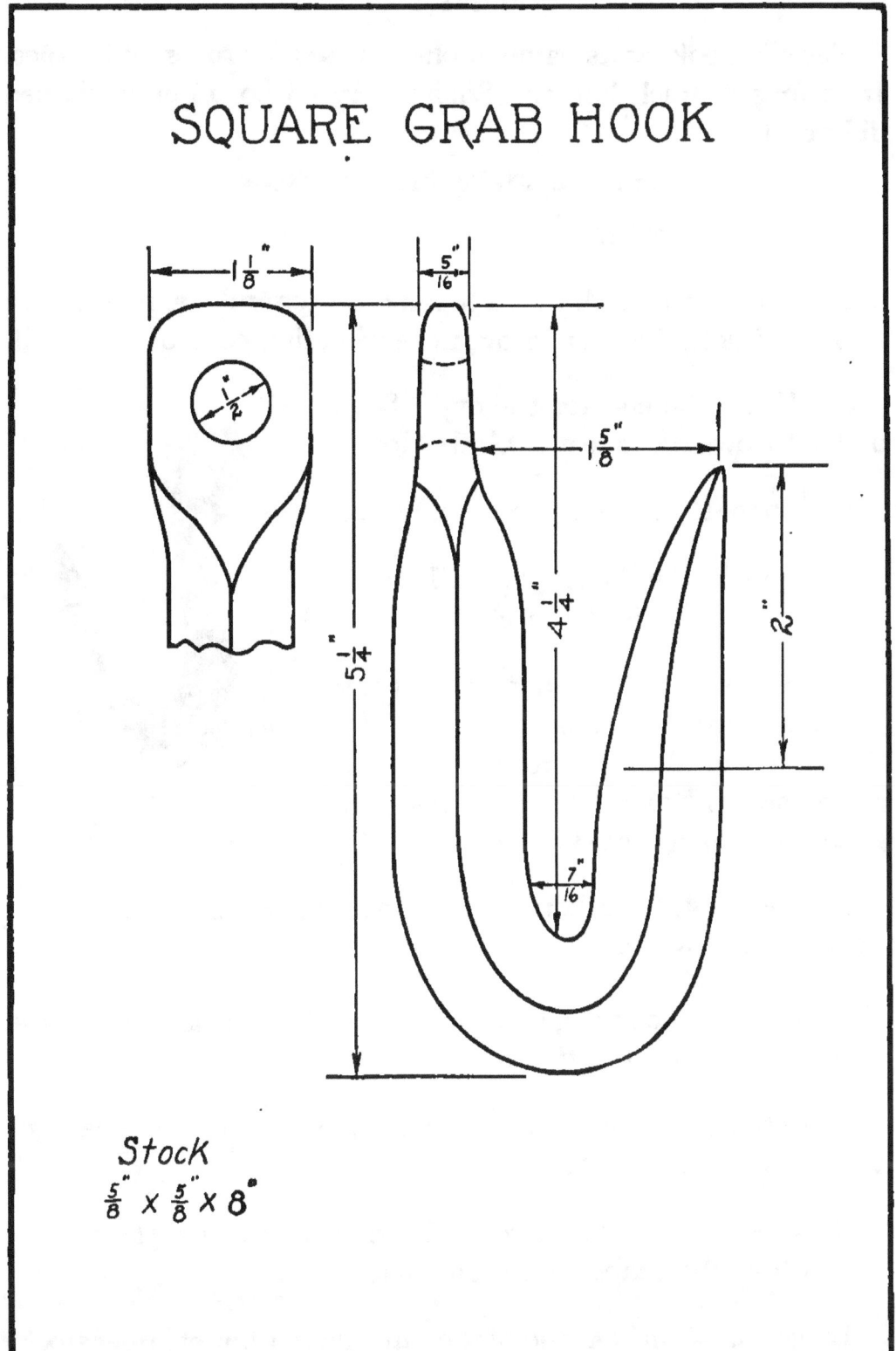

SQUARE GRAB HOOK

Stock
$\frac{5}{8}" \times \frac{5}{8}" \times 8"$

PLATE 8

SLIP HOOK

The slip hook, as its name implies, is used for ropes and cables. It is forged much like the Square Grab Hook, but is shaped differently

How to Make the Slip Hook

1.   Mark point AA.

2.   Shoulder in at AA with top and bottom fullers, if you have them.   If not, shoulder in over the rounding edge of the anvil.

3.   Upset the end for the eye.   See 1 under "How to Make the Grab Hook."

4.   Shape the head as shown on Plate 9.

5.   Punch the hole.   See "How to Punch Holes" under Gate Hinge.

6.   Shape the remainder of the hook. Use a top swage, if you have one, to make the corners on the inside rounding.   Look at the section taken thru the hook part, at the bottom of Plate 9.

FIG. 26.   SLIP HOOK.

7.   Bend over the horn of the anvil to the shape shown in Plate 9 and Fig. 26.

Notes.—The hook is made wider thru the central part, at the bottom, to make it stiff and strong.

The strain on the hook is indicated by the center line thru the eye and center of curve.

Be sure to bend the hook to the side, just below the eye, so it will take the shape shown on Plate 9.

Larger sizes can be made of the same shape, but of larger stock.

SLIP HOOK

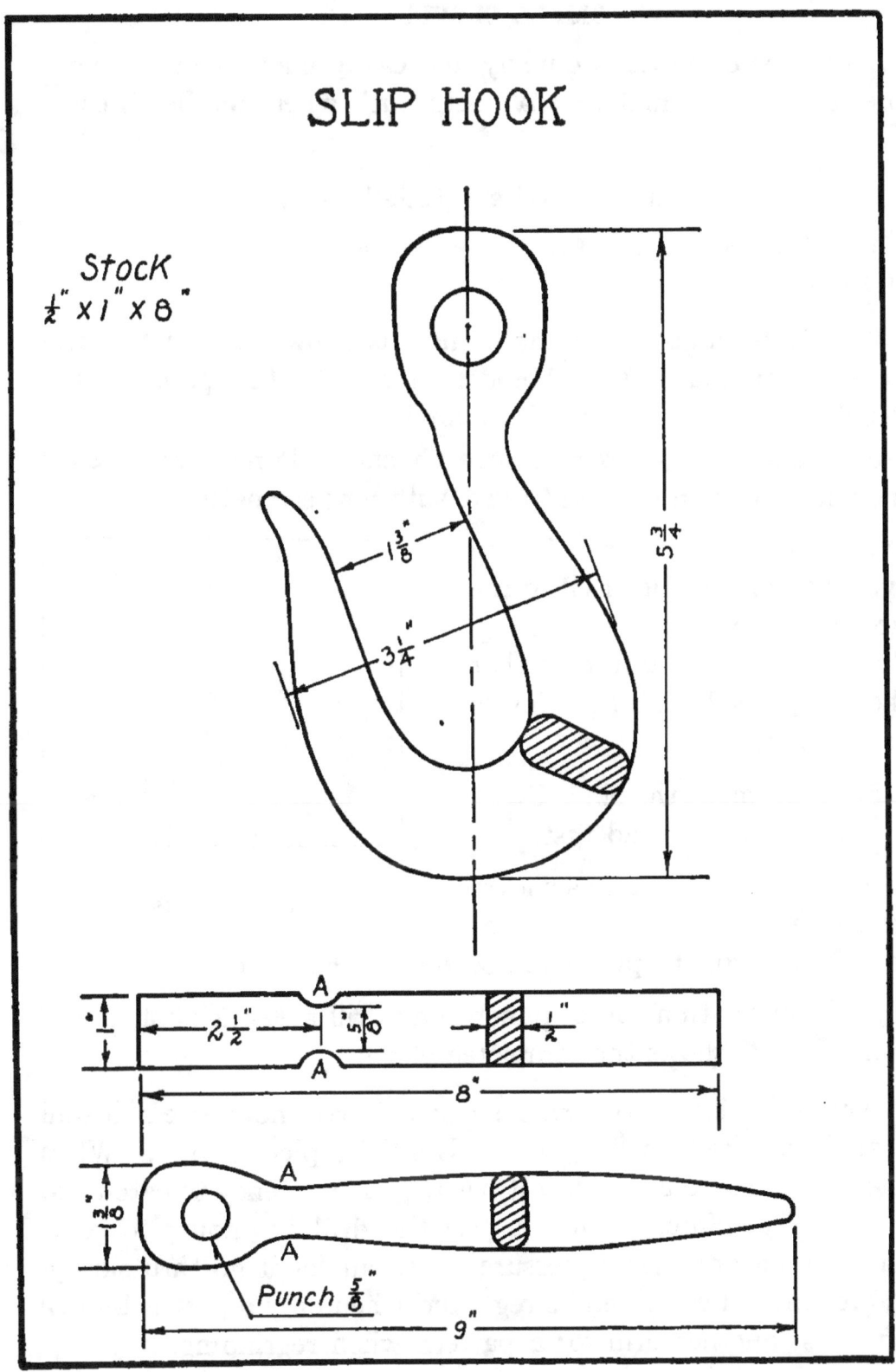

PLATE 9

SMALL CLEVIS

This type of clevis is quickly and easily made, and is, therefore, the kind to make for a "help out" when another breaks.

How to Make a Small Clevis

1.   Upset both ends of the 14-inch piece of stock for a distance of 1 inch.

2.   Flatten and shape these ends as shown in the top and front views, Plate 10.   Shoulder in for the flat part, on the rounding edge of the face of the anvil.

3.   Punch a 9/16-inch hole in each end.   Punch the hole far enough back from the end to give sufficient strength.

4.   Mark the center and bend over the horn of the anvil to the size indicated in the drawing. Be sure to bend the iron so that the straight side is on the inside of the clevis.

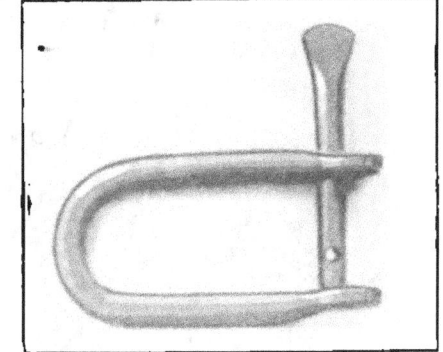

5.   Now make the pin.   Upset the stock for the head first.

6.   Shape the head as shown on the front view and the sepa-rate drawing of the pin at the bottom of the plate.

FIG. 27.
SMALL CLEVIS.

7.   Mark with a center punch, and drill a 1/4-inch hole in the end.   Fig. 27 shows the completed clevis.

Note.—Drilling thru round stock requires more care and skill than is required on flat stock.   Hold the piece firmly.   When the drill starts to come thru it has a great tendency to catch and break.   Therefore, when you feel the drill beginning to come thru do not apply any pressure to it, but let it go thru slowly. A broken drill can often be reground when only a part is broken off.   Use another drill for a pattern when regrinding.

SMALL CLEVIS

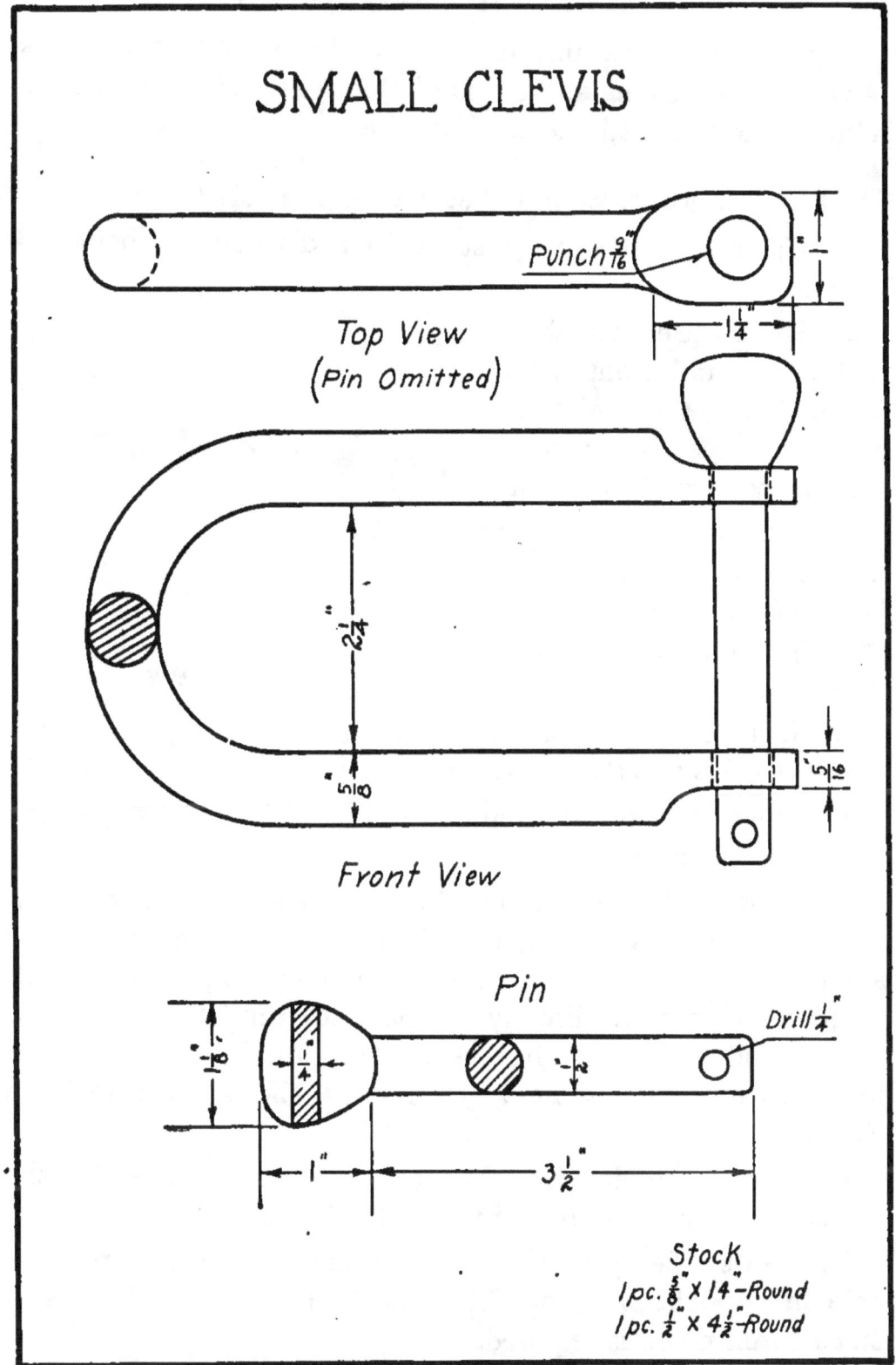

Punch $\frac{9}{16}$"

Top View
(Pin Omitted)

$2\frac{1}{4}$"

$\frac{5}{8}$"

$\frac{5}{16}$"

Front View

Pin

Drill $\frac{1}{4}$"

$1\frac{1}{8}$"

$\frac{1}{4}$"

$\frac{1}{2}$"

1"

$3\frac{1}{2}$"

Stock
1 pc. $\frac{5}{8}$" x 14"-Round
1 pc. $\frac{1}{2}$" x $4\frac{1}{2}$"-Round

PLATE 10

CLEVIS WITH THREADED PIN

When a person has sufficient time, and a set of thread plates, he will do well to make a clevis of this type. It is more satisfactory than the small clevis, Plate 10.

How to Make the Clevis with Threaded Pin

1. Upset each end of the stock for a distance of about 1½ inches.

2. Shape the ends as shown in the top and front views, Plate 11. To make the shoulder, shape over the round edge of the face of the anvil.

3. Drill or punch a $1\frac{1}{16}$-inch hole in one end, and drill a ½-inch hole in the other. Tap the ½-inch hole with a ⅝-inch tap.

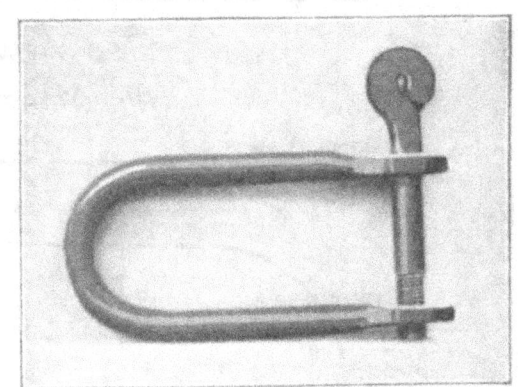

FIG. 28.
THREADED PIN.

4. Bend over the horn of the anvil to the proper size, and with the holes exactly above each other.

5. Make the pin. Flatten one end to $\frac{5}{16}$ inch thick, ⅝ inch wide and 2¼ inches long.

6. Heat, and clamp in the vise, and bend the flattened end over to one side, as in Fig. 16. Then bend over the horn of the anvil until the curve for the head is almost completed. Finish by hammering down directly on the stock until the end comes around and touches. The eye of the gate hook, Plate 3, is forged in much the same way. Fig. 28 shows the completed clevis.

7. Thread the pin with a ⅝-inch die. Be sure the end of the pin is perfectly round before trying to thread it.

Note.—Use plenty of oil when threading a bolt or tapping a hole in wrought iron, or "blacksmith iron" or "mild steel." In cast iron no oil is required.

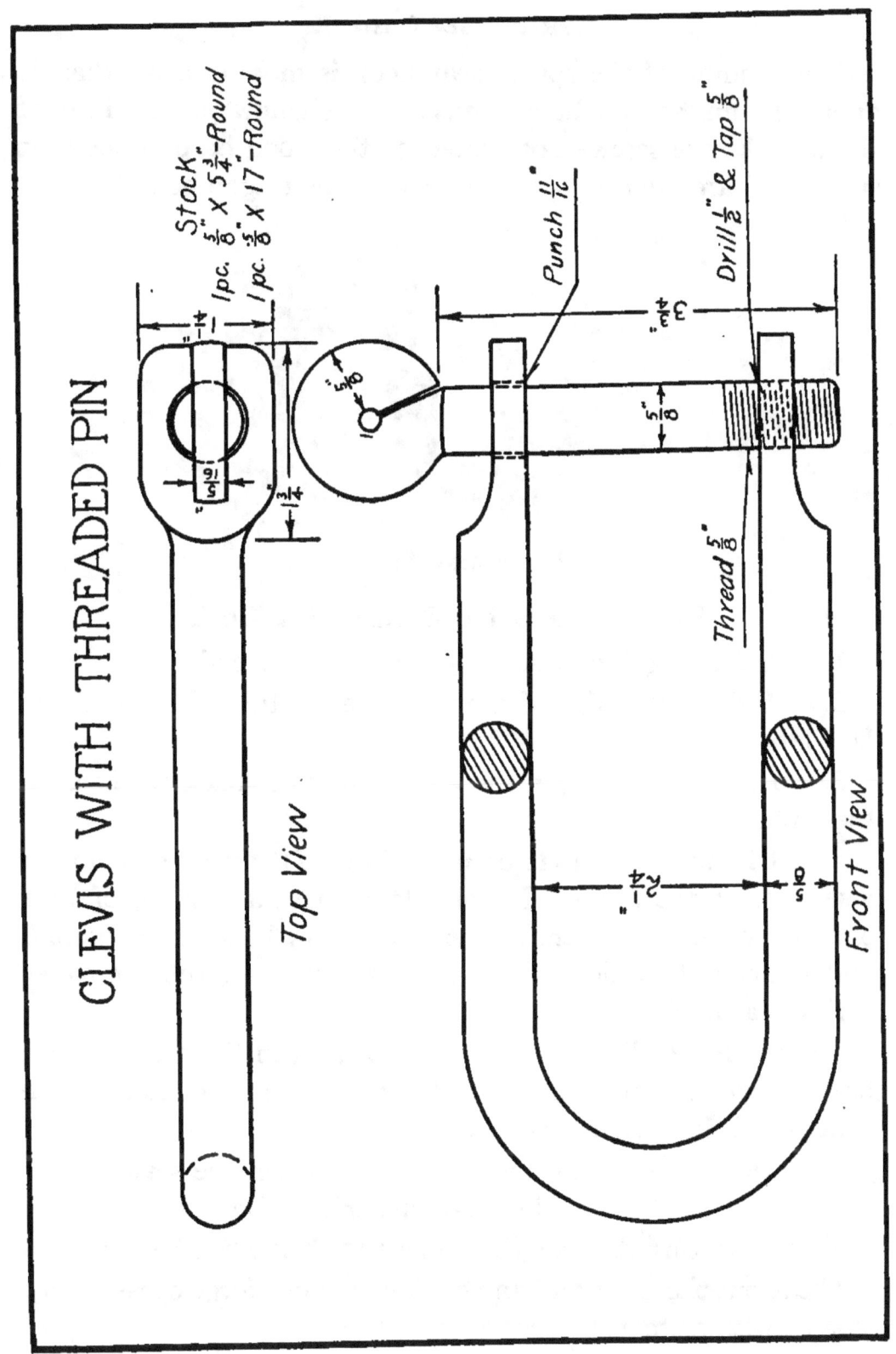

CLEVIS WITH THREADED PIN

Top View

Front View

Stock
1 pc. $\frac{5}{8}$" X $5\frac{3}{4}$"-Round
1 pc. $\frac{5}{8}$" X 17"-Round

Punch $\frac{11}{16}$"

Drill $\frac{1}{2}$" & Tap $\frac{5}{8}$"

Thread $\frac{5}{8}$"

PLATE 11

SPRING SEAT HOOK

The bending of the spring seat hook is more difficult than is apparent, unless you have a vise with shallow jaws. Two of these hooks are screwed or bolted to the wood base of the seat spring and are used to hang the seat on the wagon box.

Fig. 29.   Spring Seat Hook.

How to Make the Spring Seat Hook

1.   Drill the ¼-inch hole 1½ inch from one end.

2.   Mark on the edge of the iron, the location of A, B, C and D, Plate 12.

3.   Bend at A and B, in the vise or over the back or "heel" of the anvil.

4.   It is not likely that corners C and D can be bent in the vise or over the anvil. If not, clamp a square piece of iron, one inch or less in width, in the vise. Now bend the corners C and D about this piece. Fig. 29 is a picture of the completed spring seat hook.

Notes.—In bending about a piece of iron in the vise you will have to hammer first on one side of a corner and then on the other until the corner is square.

Reduce the 1-inch distance to ⅞ inch when the seat is to be used on a wagon box made of 1-inch surfaced boards.

Be sure to turn flat iron over now and then while heating.

Clean out the ashes and any clinkers in your forge occasionally, just as with a stove.

SPRING SEAT HOOK

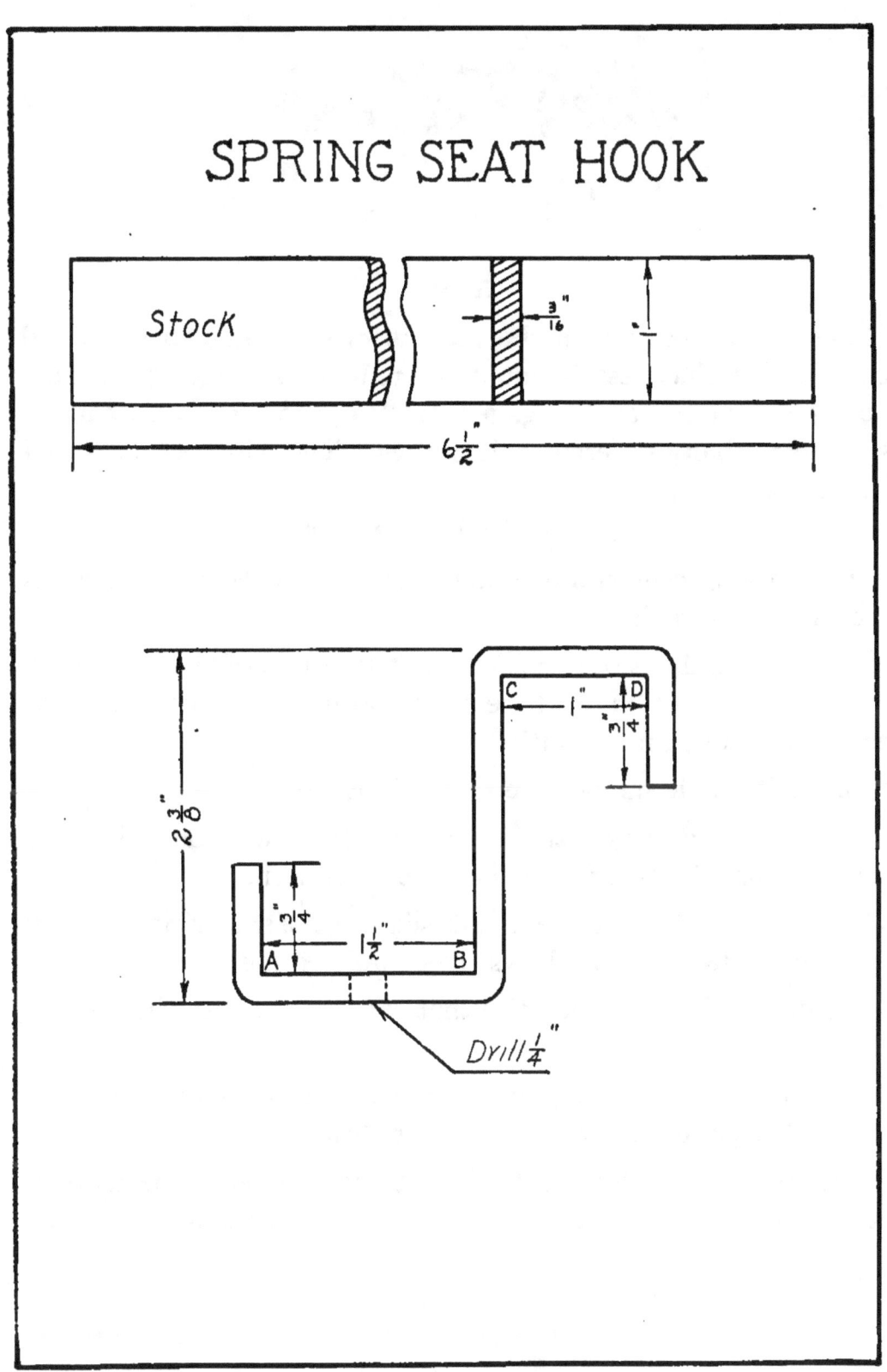

PLATE 12

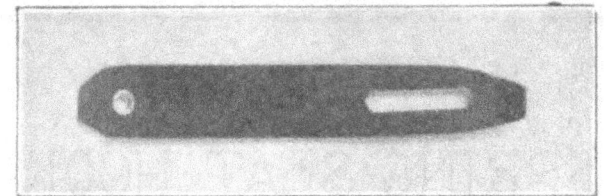

Fig. 30. Hasp.

HASP

The stock from which the hasp is made is known as "band iron." The thickness is not measured in fractions of an inch, but is designated as "12-gage" or "14-gage," etc. It can be secured in many different thicknesses. The width is measured in inches.

How to Make the Hasp

1. Locate, center-punch and drill three ⅜-inch holes as indicated on drawing.

2. Heat and cut out the stock between two of the holes with a hot chisel. Cut on the base of the horn of the anvil. Do not cut to the full ⅜-inch width.

3. With a file smooth the sides of the slot.

4. Heat and cut off the corners to the dimensions shown in the drawing. Then smooth the edges with a file.

5. Heat and bend the point slightly as shown in the top view, Plate 13. Fig. 30 shows the completed hasp.

Notes.—This hasp is of a size that can be used with the staples, Plate 2.

The holes in the hasp may be punched, but on such thin stock as this better results are obtained by drilling.

The cast iron swage block, Fig. 7, can frequently be used to advantage in shaping articles out of sheet or band iron.

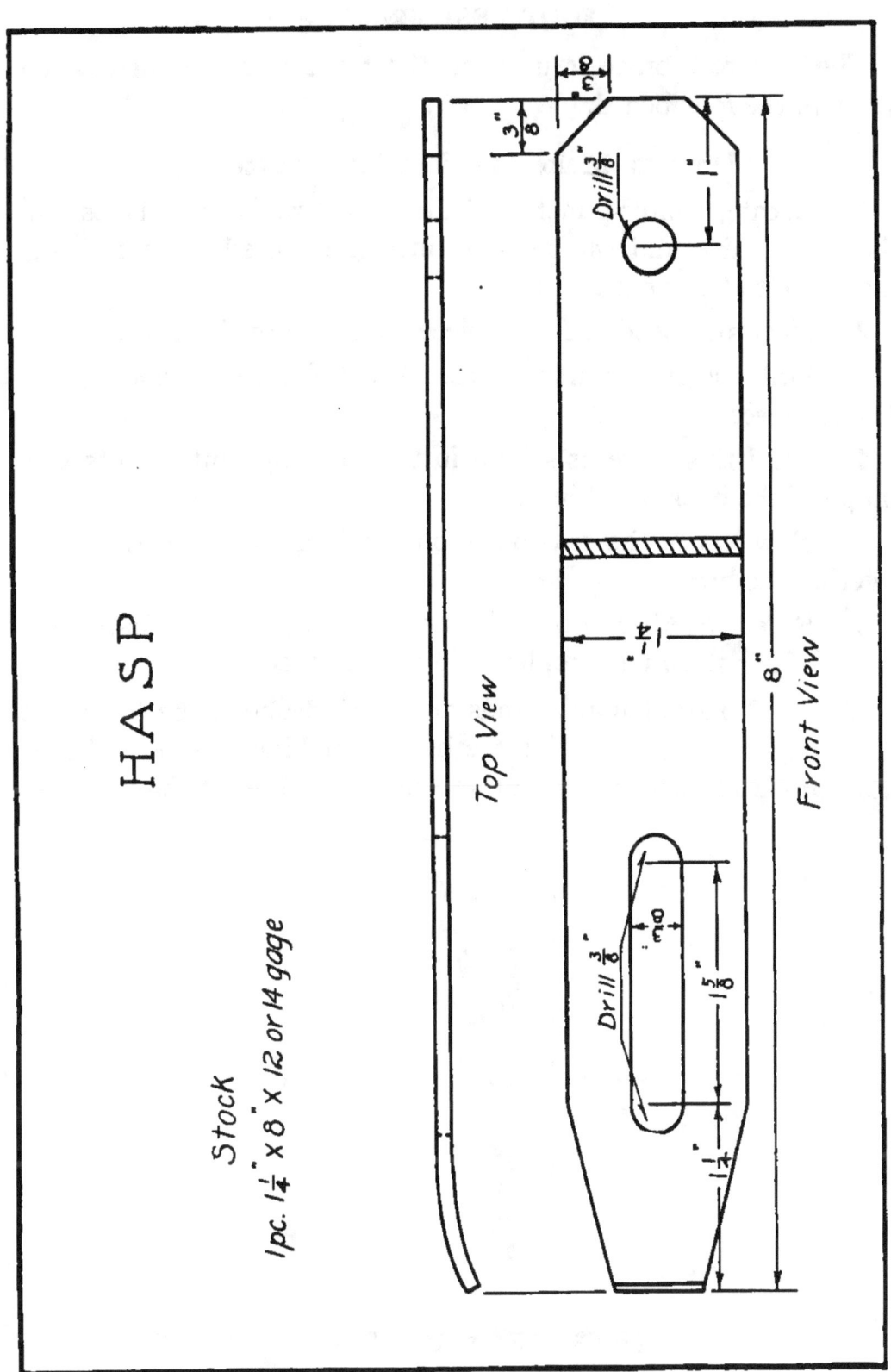

HASP

Stock
1pc. 1¼" x 8" x 12 or 14 gage

Top View

Front View

Drill ⅜"

Drill ⅜"

PLATE 13

FOOT REST BRACE

The foot rest brace is used on the front of a wagon box to support the foot board.

How to Make the Foot Rest Brace

1. Locate, center-punch and drill the two ¼-inch holes for the foot board, and the three ¼-inch holes used for attaching the brace to the wagon box.

2. Heat and bend the long piece as shown in Plate 14.

3. Drill the two ¼-inch holes. ½ inch from each end of the 13-inch piece.

4. Bend this piece as shown in the drawing, until it fits the shape of the back and top parts.

5. Now locate the two holes on the back and top used for riveting the brace.

6. Rivet together with a ½-inch rivet of ¼-inch diameter. In Fig. 31 is shown a completed foot rest brace.

Note.—Braces of many kinds are needed about the farm for one thing or another. This method of making one out of flat iron can be applied to many other braces of different shapes.

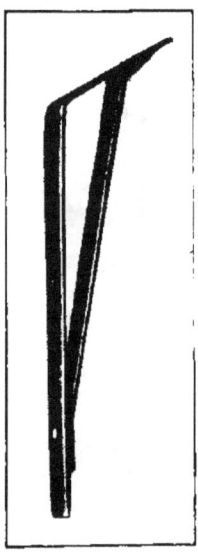

Fig. 31. Foot Rest Brace.

FOOT REST BRACE

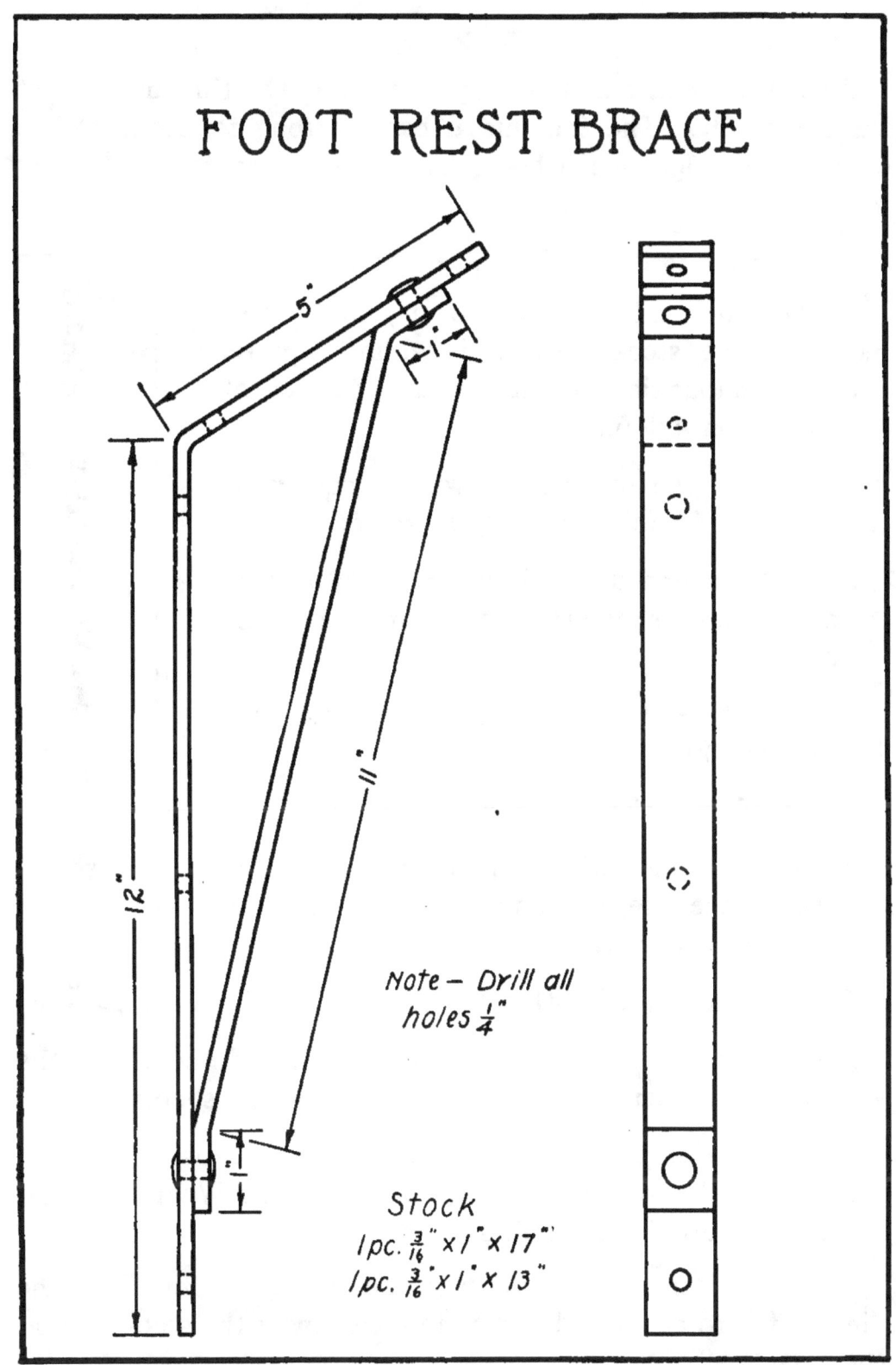

Note — Drill all holes $\frac{1}{4}$"

Stock
1 pc. $\frac{3}{16}$" x 1" x 17"
1 pc. $\frac{3}{16}$" x 1" x 13"

PLATE 14

STRAP BOLT

Considerable care is necessary in shaping the threaded end of the strap bolt. The thin metal has a strong tendency to lap over on the edges when being hammered down to the ⅜-inch round.

How to Make the Strap Bolt

1. On one end hammer down about three inches of the stock until it is almost square. Hammer on four sides, when necessary, to prevent the edges from folding over.

2. Round the end just drawn out square so that it will be like the drawing, Plate 15.

3. If the rounded part is less than ⅜ inch in diameter, upset as much as is necessary and round the end again.

4. Drill three ¼-inch holes as shown in Plate 15 and Fig. 32.

5. Bend A in the vise or over the anvil.

6. Bend B around a square piece of iron clamped in the vise, just as in 4, under "How to Make a Spring Seat Hook," Plate 12.

7. Thread the end with a ⅜-inch die.

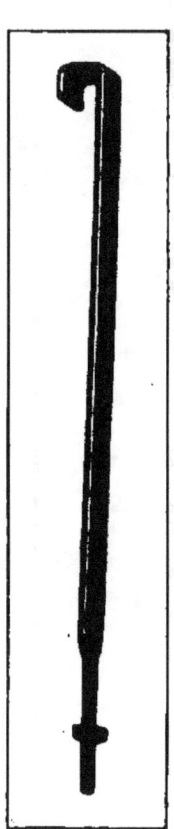

Fig. 32.
Strap Bolt.

Notes.—Always measure the height of the side boards of the wagon box to determine how long the strap bolts should be.

Should the edges fold over when you are drawing the end round bring the iron to a welding heat (see Lap Weld, Plate 18, for welding) and hammer all around the end.

The end of the round part must be absolutely round, or the die will fail to catch and will simply cut away the stock around the outside.

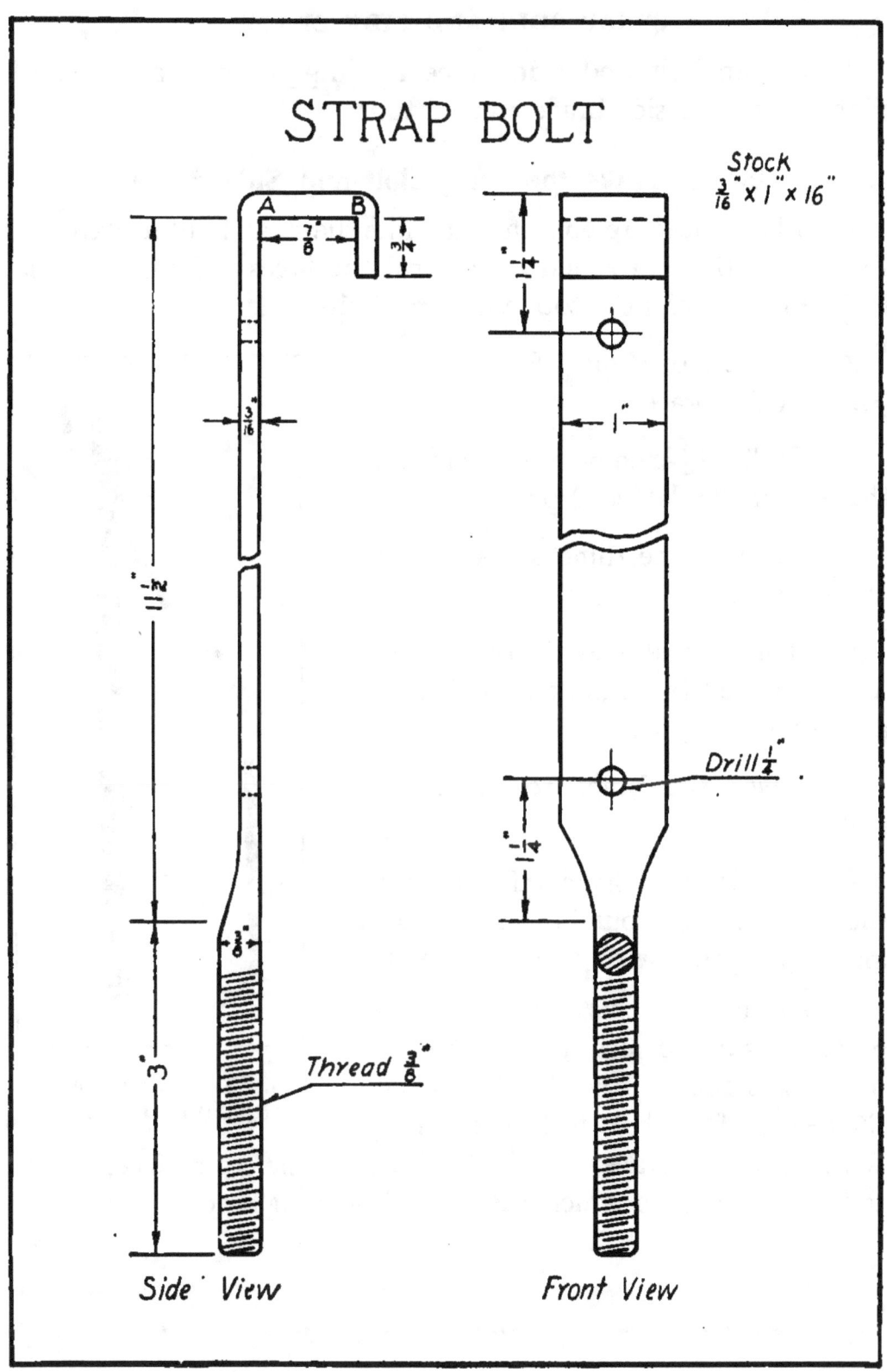

STRAP BOLT

Stock
$\frac{3}{16}$" x 1" x 16"

A   B

Side View   Front View

Drill $\frac{1}{4}$"

Thread $\frac{3}{8}$"

PLATE 15

STRAP BOLT AND SIDE BRACE

This strap bolt and side brace is simply the strap bolt of Plate 15 with a side brace riveted to it.

How to Make the Strap Bolt and Side Brace

1. Make the strap bolt part just like the strap bolt of Plate 15. First draw the end round; then drill the holes adding the one for the rivet; bend the top, and thread the end.

2. Flatten one end of the round piece for the brace.

3. Drill a ¼-inch hole ½ inch from the end of the flattened part.

4. Bend to the dimensions shown in the drawing.

5. Thread the end of the brace. You will need two nuts on the brace, as shown in Fig. 33.

6. Rivet the brace to the strap bolt.

Notes.—Be just as careful to get the ends perfectly round before threading as you were with the strap bolt.

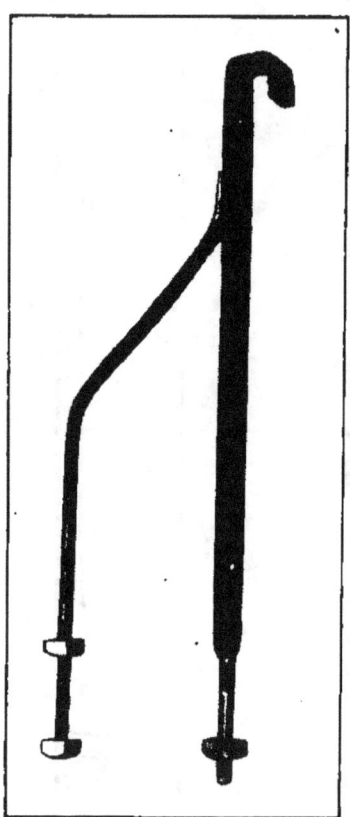

FIG. 33. STRAP BOLT AND SIDE BRACE.

When making this forging for a repair be sure to measure the height of the side boards, the distance from the center of the bolt to the center of the hole for the brace, and the thickness of the cross-piece, under the wagon box, to which the strap bolt and brace are bolted.

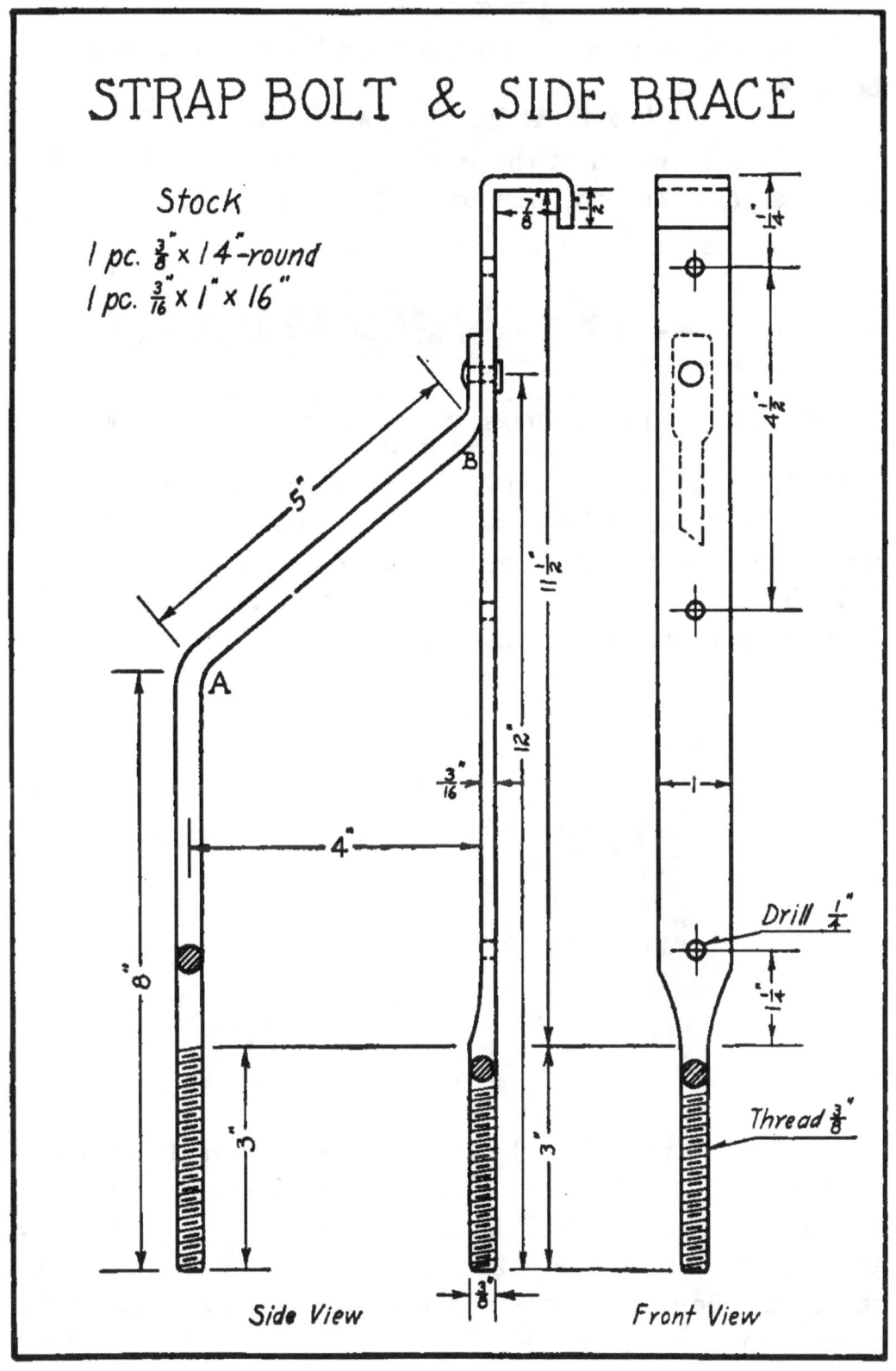

STRAP BOLT & SIDE BRACE

Stock

1 pc. $\frac{3}{8}" \times 14"$-round
1 pc. $\frac{3}{16}" \times 1" \times 16"$

Side View

Front View

PLATE 16

JOIST STIRRUP

The making of a joist stirrup is a good test of your ability to bend iron

How to Make the Joist Stirrup

1. Find the center of the stock and mark points A, B, B, A on the edge of the iron. See top of Plate 17.

FIG. 34. CENTER HAMMERED TO ONE SIDE; ENDS HAMMERED DOWN AT ARROWS.

2. Heat and place in the vise at one of the A points and with a tongs twist the end a quarter turn. Hold the iron with the tongs about two inches above the top of the vise.

3. Twist above the other point A in the same manner, being sure that you twist in the right direction

FIG. 35   HAMMERING DOWN THE CENTER.

4. You will note that the twists are now thru the center of the stock. The central part, Fig. 34, must all be hammered over to one side, as to the line E-E; and the ends hammered down as shown by the arrows. Fig. 35 shows the method of hammering down one side of the center. The outside ends are hammered down in the same manner.

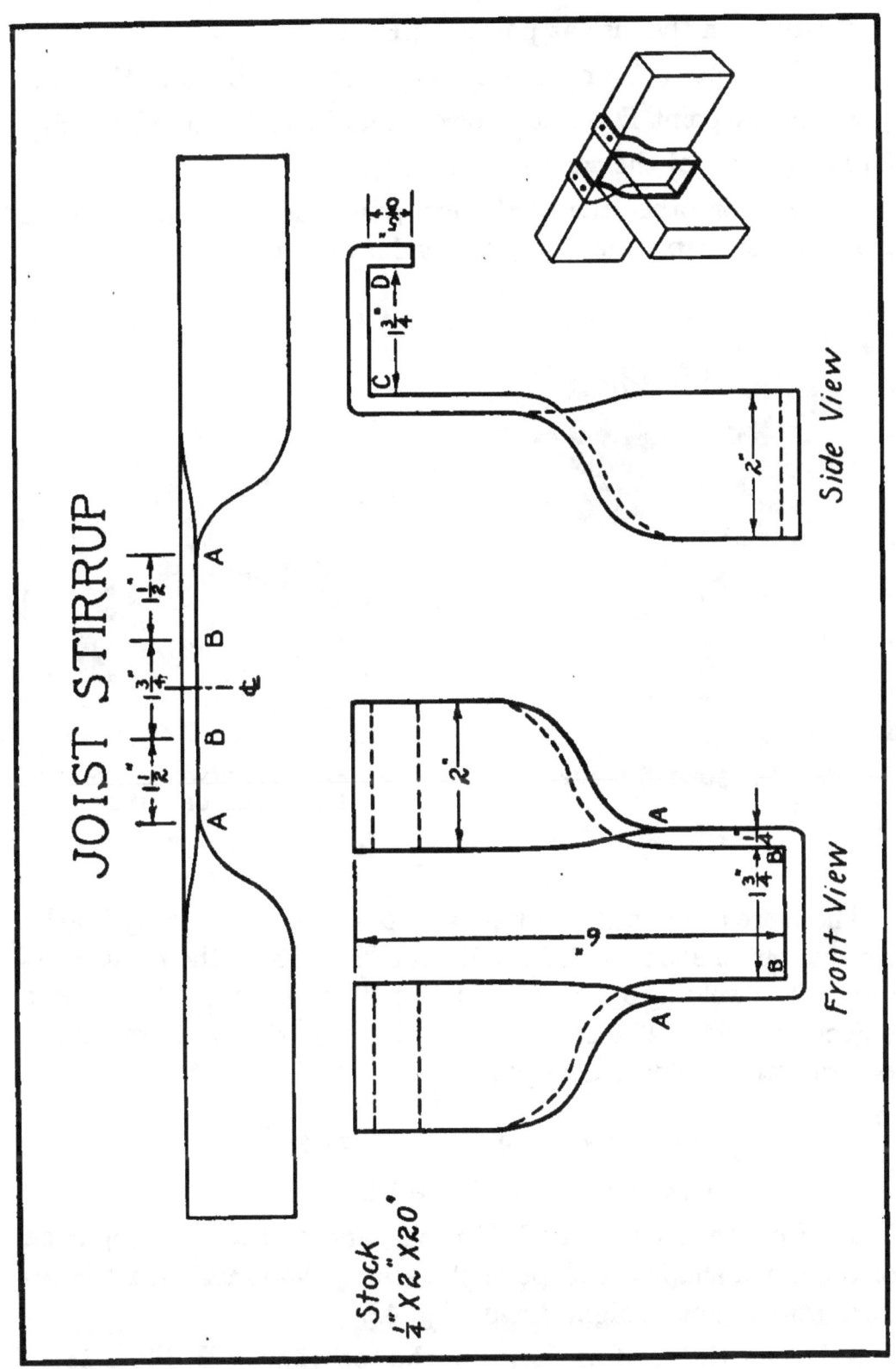

JOIST STIRRUP

Stock ¼" X 2" X 20"

Front View

Side View

PLATE 17

5. Bend in the vise at points B, B.

6. Mark point C on both upright parts and bend in the vise.

7. Mark point D on both pieces and bend in the vise. Fig. 35 is a picture of the completed stirrup.

Note.— For other than 2x6-inch joists, add enough stock to make the stirrup either longer or wider or both.

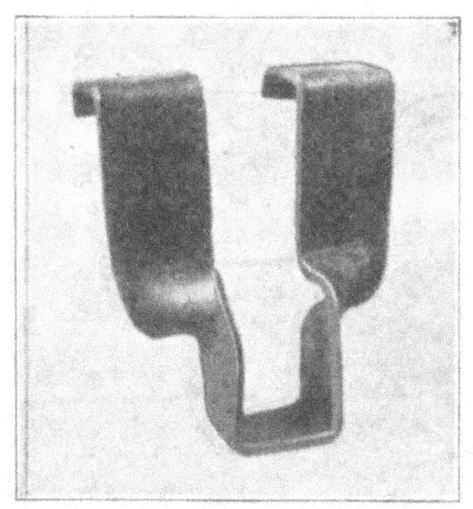

FIG. 36. JOIST STIRRUP.

FIG. 37. SHOWING SCARFS ON LAP WELD OF RING.

LAP WELD

There are a number of different types of welds used by blacksmiths, but the lap weld, Fig. 37 and Plate 18, is the most common. Try welding a ring, C, Plate 18, first, as that is the easiest object to weld. The scarfs can be bent to the position shown before actual welding is begun.

How to Shape the Scarfs

1. Upset both ends, see A, Plate 18.

2. Hold the iron with a bolt tongs, and on the rounding edge of the anvil, shape the slope of the scarf. Hammer with a *backward* stroke, not straight down, Fig. 42.

3. Point the scarf as shown in the top view of B, Plate 18.

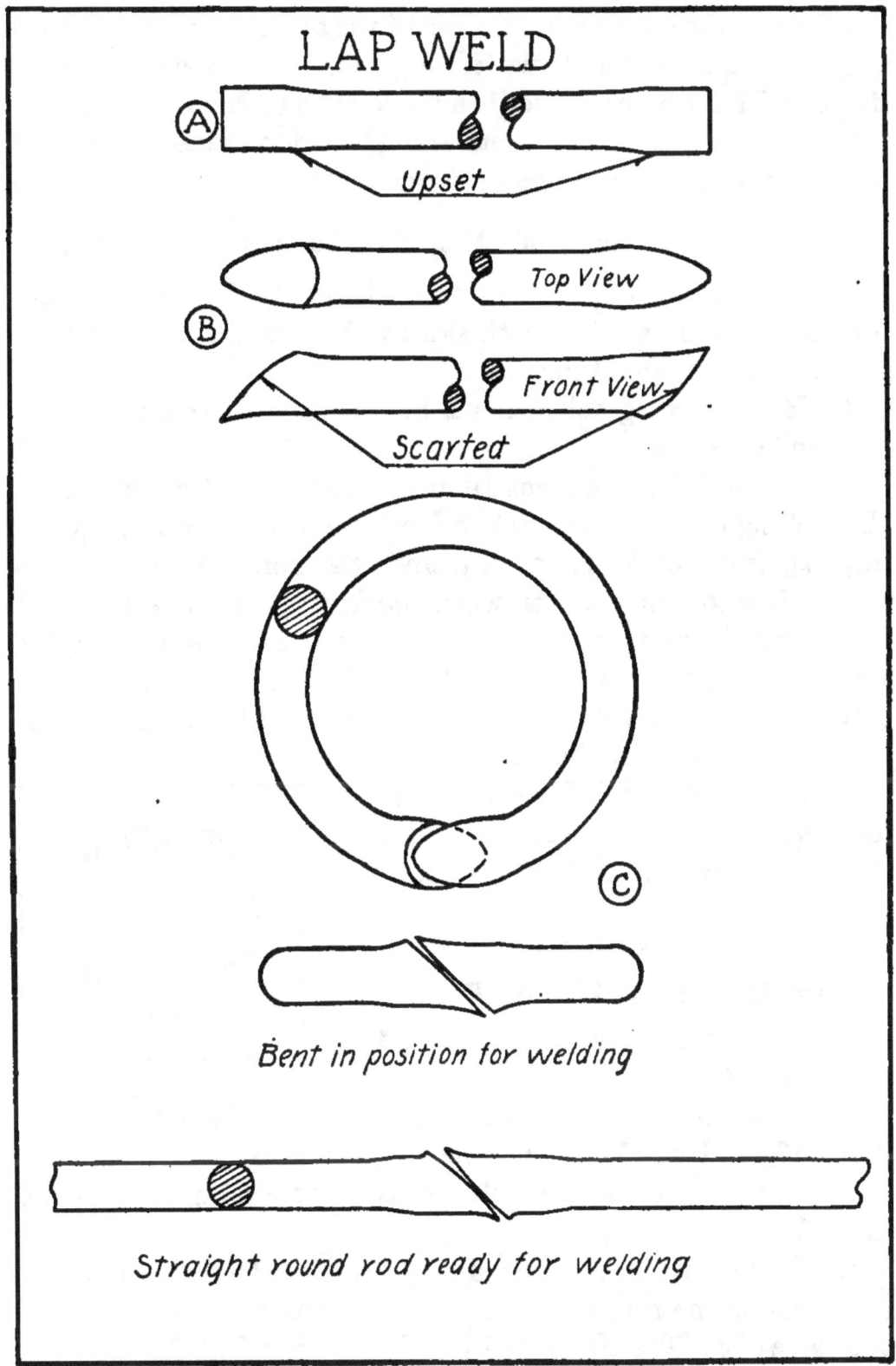

LAP WELD

A Upset

B Top View

Front View

Scarfed

C

Bent in position for welding

Straight round rod ready for welding

PLATE 18

4.   Bend the scarf over the horn of the anvil to make the slope slightly rounding, as shown in Fig. 43.   The scarfs must be shaped with a convex curve if the weld is to be successful.

5.   The length of the scarf should seldom exceed one and one-half times the thickness or diameter of the iron.

Always observe the following directions when welding:

1.   Use little draft in your fire.   Heat the iron slowly and turn occasionally so that both sides will reach the welding temperature at the same time.

2.   Have burning coke above, below and to the sides of the iron while heating.

3.   The welding heat has been reached when the iron looks like melting ice, and appears to be melting on the surface.   When tiny explosive sparks shoot off from it the iron is *burning*.   It is not too late to weld the iron when the first sparks appear.

4.   Quickly remove the iron from the fire to the face of the anvil, place the two scarfs together and strike a couple of light blows on the weld, followed by several heavy blows.   Strike the weld on both sides of the iron.

5.   Shape the welded part to conform with the shape of the iron, and then shape the object, if that is necessary.

FIG. 38.
WELDED RING.

WELDED RING

How to Make a Welded Ring

1.   Read and study the Lap Weld and Plate 18 first.

2.   Upset both ends of the stock, Plate 19, and scarf the ends like B, Plate 18.

3.   Bend in a circle and into position for welding, and make the weld.

4.   Shape the iron round and to size at the place of welding.

5.   Shape the ring over the horn of the anvil, or a mandrel if you have one.   The completed ring is shown in Fig. 38.

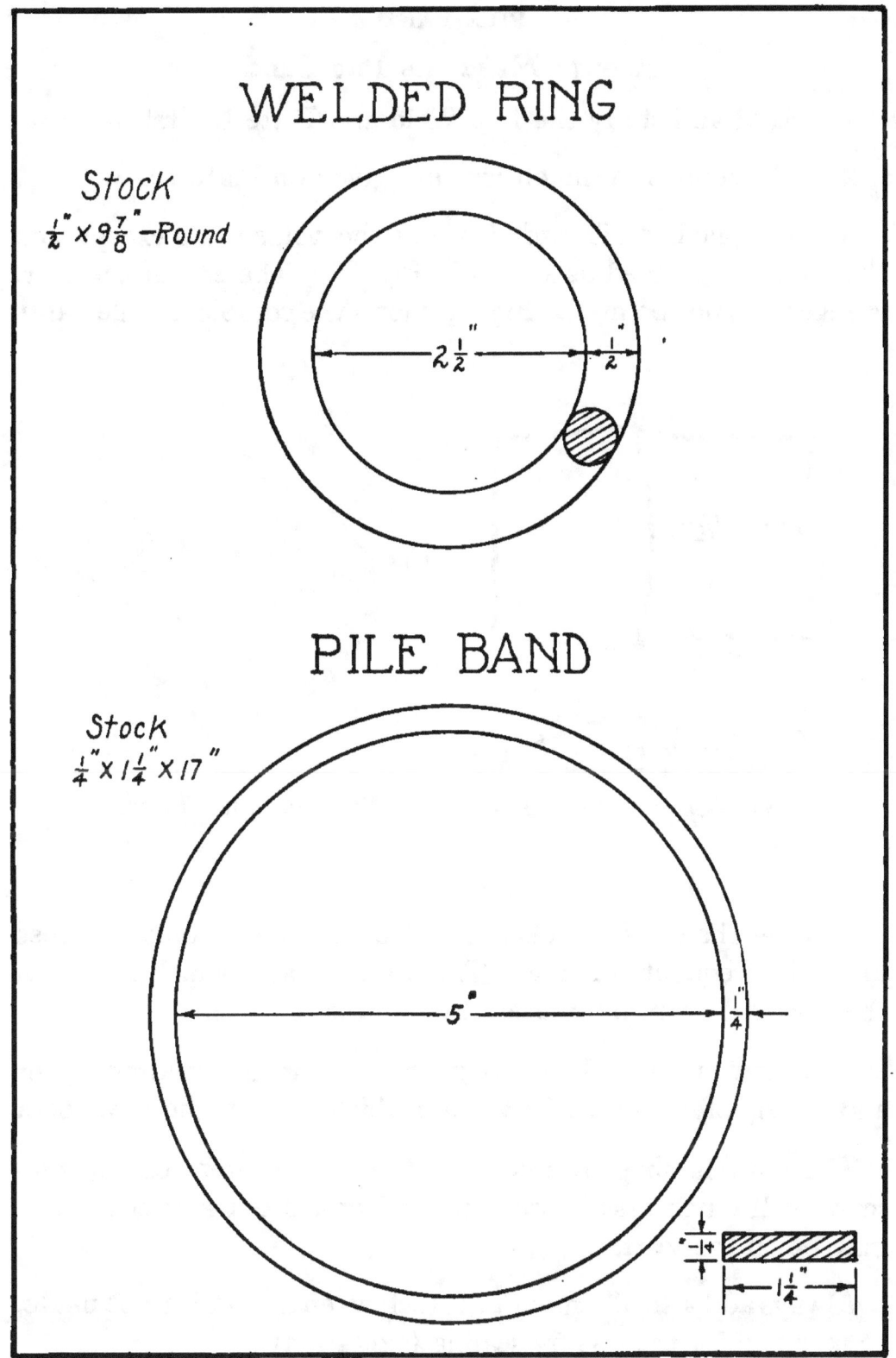

WELDED RING

Stock
½" × 9⅞" — Round

2½"   ½"

PILE BAND

Stock
¼" × 1¼" × 17"

5"   ¼"

¼"   1¼"

PLATE 19

PILE BAND

How to Make the Pile Band

1. Read and study the Lap Weld and Plate 18 first.

2. The stock and dimensions are given on Plate 19.

3. The pile band is made just like the welded ring except that the scarfs are shaped as shown in Fig. 39. The end of the scarf is edged but not pointed. Fig. 40 shows the completed pile band.

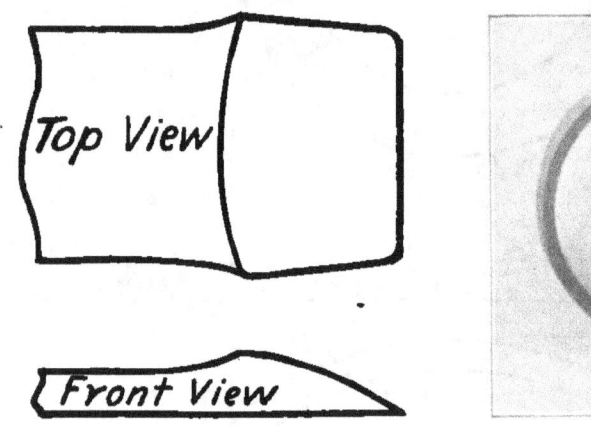

FIG. 39. SCARF OF PILE BAND.

FIG. 40. PILE BAND.

Notes.—The scarfed ends are first upset to overcome the loss in size by frequent heating. The scale (oxide) which forms on the iron and drops off, materially reduces the size.

The scarf is pointed, on round iron to make welding easier and to help prevent unsightly edges which are not wholly welded.

The scarf is shaped curved so that the centers of the two scarfs will touch first in welding and crowd out any scale that may be on the scarf.

Always add a small amount of stock when a weld is to be made, to make up for the loss by scaling (oxidation).

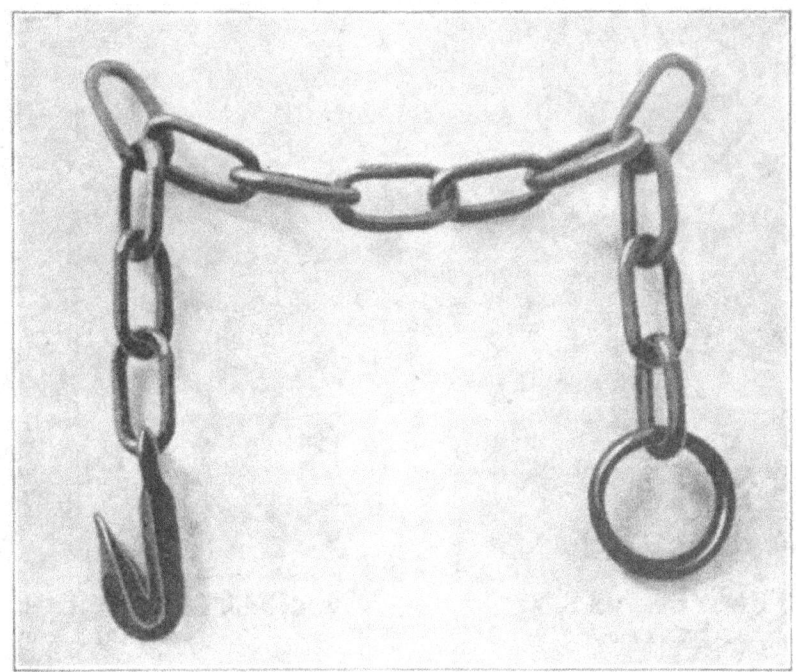

FIG. 41. CHAIN WITH RING AND GRAB HOOK.

CHAIN LINKS

The making of chain links is intended almost wholly for making repairs, tho a whole chain may be made, Fig. 41.

How to Make a Chain Link

1. Read and study the Lap Weld and Plate 18 first.

2. Upset the ends as in 2', Plate 20.

3. Bend over the horn of the anvil to the shape indicated in 3.

4. Scarf the ends as shown in 3. They are shaped just as for the welded ring. See Figs. 42 and 43. Remember to make the scarfs on *opposite* sides of the two prongs of the link.

5. Shape the link as shown in 4, Plate 20.

6. Weld and shape as in 5.

7. Figure out the stock necessary for this link. See Plate 1 and directions. Add about ⅛ inch to the length to make up for the loss in welding and scaling.

FIG. 42. FIRST STEP IN SHAPING
A SCARF.

FIG. 43.   BENDING POINT OF
SCARF.

How to Make a Chain

1.   Weld two links separately.

2.   Weld these two together with a third.

3   Weld three more in the same manner.

4.   Weld the two groups of three together with a seventh.

5.   When a large number of links are to be made do all the shaping first and then the welding.

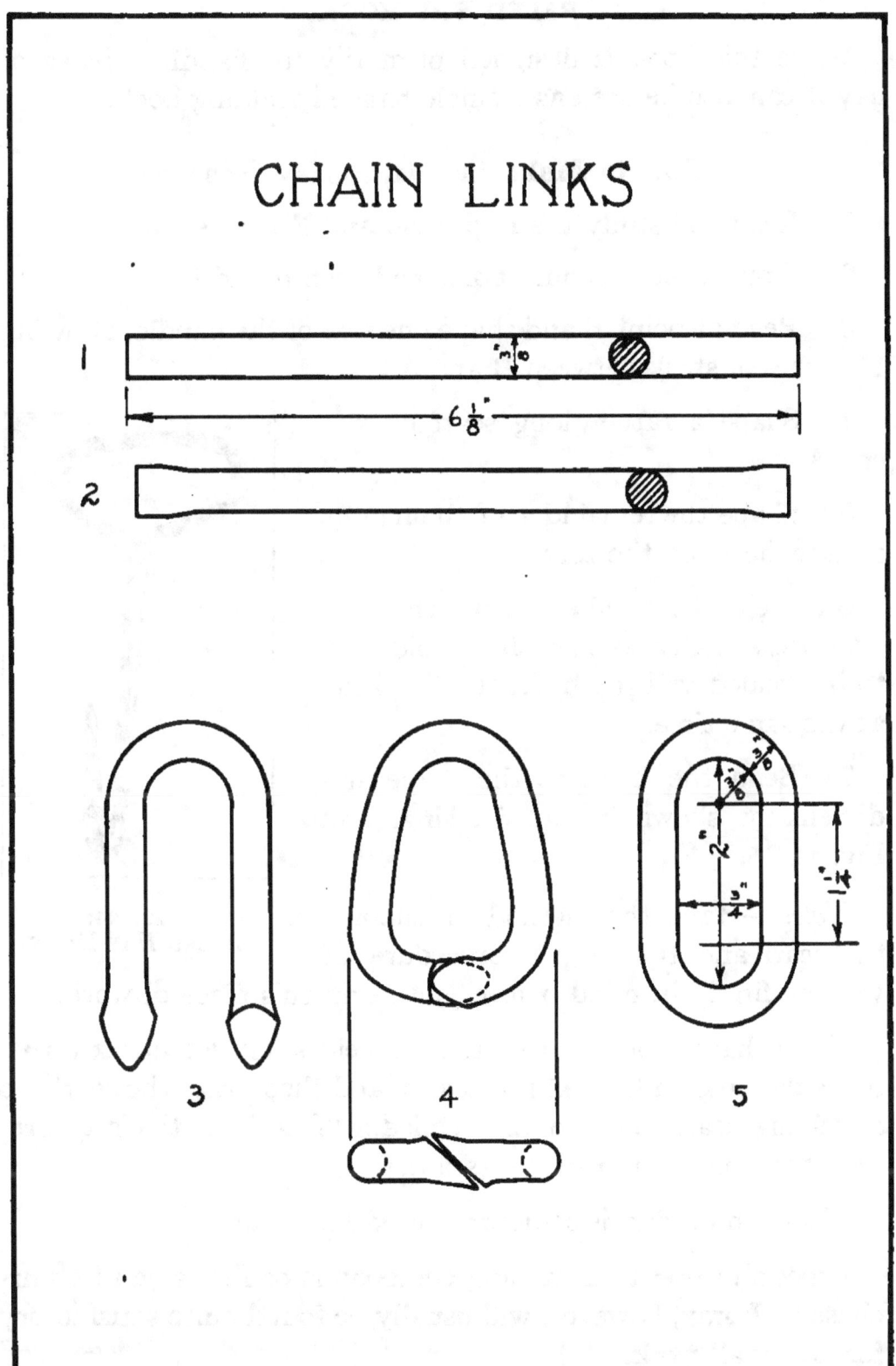

CHAIN LINKS

PLATE 20

BALED HAY HOOK

While this hook is designed primarily for handling bales of hay it can also be used as a single-handed scalding hook.

How to Make the Baled Hay Hook

1. Read and study the Lap Weld and Plate 18 first.

2. Draw a blunt square point and then round it.

3. Bend at point B and shape one side of the handle, allowing 11 inches of stock between B and A

4. Shape a rather long scarf on the end A.

5. Shape the remainder of the handle on the horn of the anvil.

6. Weld the handle. Turn the iron over occasionally so that the whole part to be welded will reach the welding heat at the same time.

7. Bend the hook to the shape and dimensions shown in the drawing, and Fig. 44.

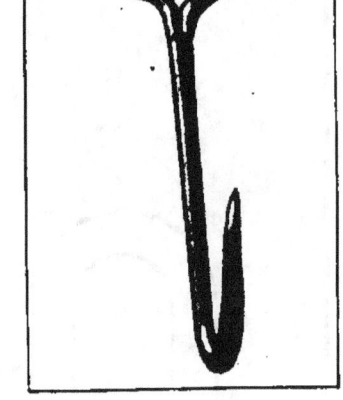

FIG. 44.
BALED HAY HOOK.

Notes.—Have the method of shaping the scarf and the proper procedure of welding firmly in mind before you begin this piece of work.

If you have trouble in making a weld stick use borax on it. Sprinkle some on the red hot scarfs, and then bring the scarfs to a welding heat. The borax is called a "flux." It tends to prevent the iron from oxidizing (scaling).

Clean white sand is sometimes used for a flux.

Especially prepared welding compounds or fluxes can be purchased. Borax, however, will usually be found quite satisfactory for nearly all work.

BALED HAY HOOK

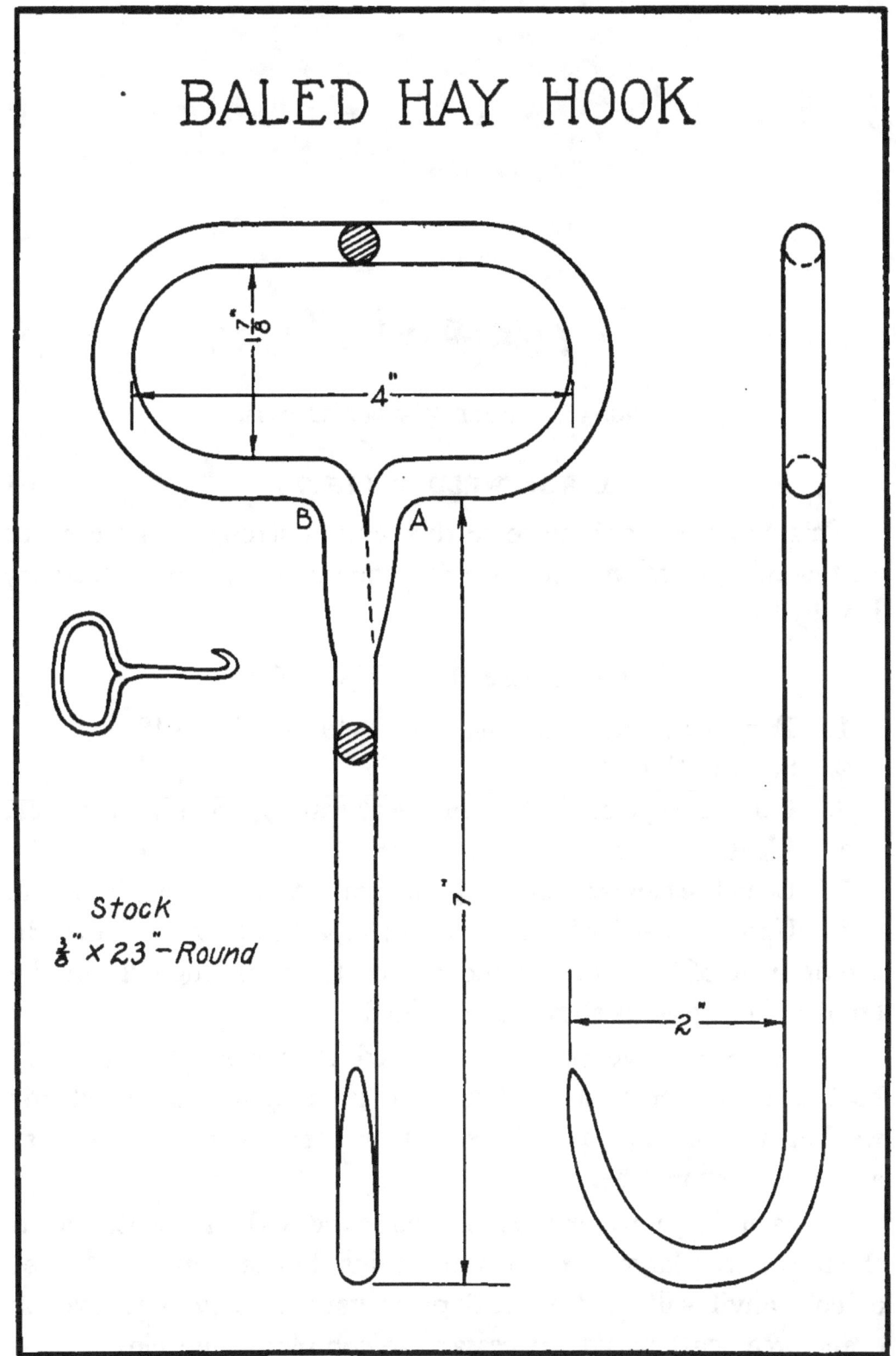

Stock
⅜" × 23" — Round

PLATE 21

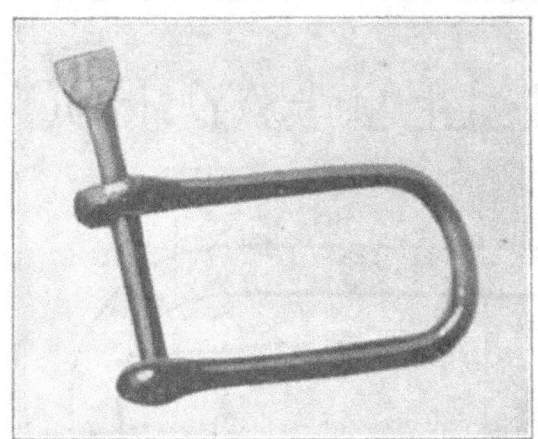

FIG. 45. LARGE WELDED CLEVIS.

LARGE WELDED CLEVIS

This clevis is made quite readily and is strong. The pin has no fastening whatever, and where frequent changes are necessary is very handy.

How to Make the Welded Clevis

1. First read and study the Lap Weld and Plate 18.
2. Scarf both ends.
3. Bend the eye so that it will be at least ¾ inch in diameter.
4. Weld.
5. Bend the clevis to the shape shown in the drawing, Plate 22.
6. Upset the end of the pin and shape the head as shown in the drawing of the pin. Shoulder in on the rounding edge of the anvil. In Fig. 45 is shown the welded clevis.

Notes.—You have probably noticed that iron at a welding heat is sticky or pasty. Therefore in shaping an object for welding be sure to have the scarfs or parts to be welded just where they should be.

Welds on large iron can usually be made with one heat, due to the fact that a large piece of iron retains its heat longer. A cool or cold anvil will chill a small piece very quickly, and two or even three heats may be necessary to complete the weld.

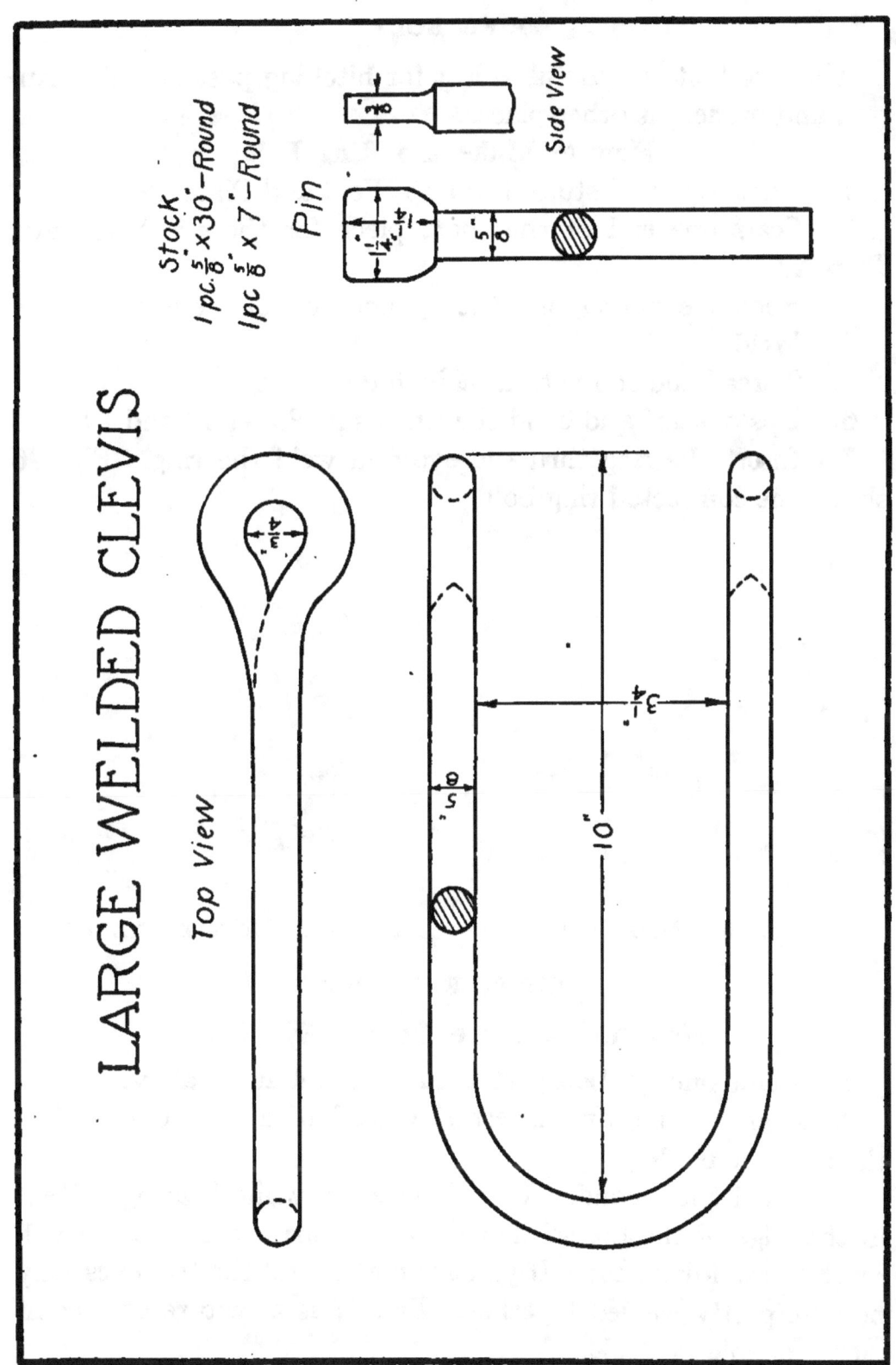

LARGE WELDED CLEVIS

Top View

Side View

Pin

Stock
1 pc. ⅝" x 30"–Round
1 pc ⅝" x 7"–Round

10"

3¼"

⅝"

1¼"

¼"

⁵⁄₁₆"

PLATE 22

RING BOLT

The ring bolt is a useful object for hitching post or rail, manger, and numerous other places.

How to Make the Ring Bolt

1. First read and study the Lap Weld and Plate 18.
2. Scarf one end of the short piece for the eye bolt part, Plate 23.
3. Bend the end to form the ⅝-inch hole of the eye.
4. Weld.
5. Thread the end with a ⅜-inch die.
6. Upset, scarf and bend the ring as in Plates 18 and 19.
7. Insert the ring thru the eye and weld the ring. Fig. 46 shows the completed ring bolt.

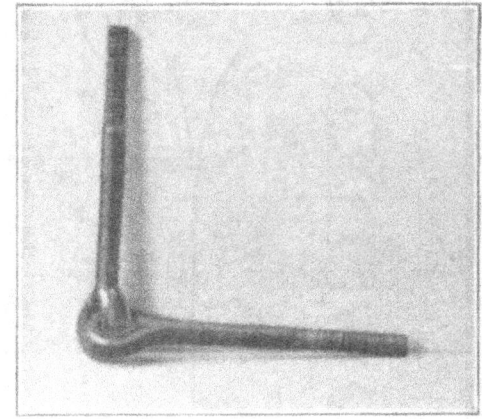

FIG. 46.   RING BOLT.   FIG. 47.   DOUBLE EYE BOLT.

DOUBLE EYE BOLT

How to Make the Double Eye Bolt

1. Make one eye bolt just as in 1, 2, 3, 4 and 5 above.
2. Scarf, shape, and insert the head of a second one thru the first one made.
3. Weld the second. Turn frequently while heating. Heat in the edge of the fire so that only the parts to be welded will reach the welding heat. If you are not careful the two eyes may become partly welded together. Fig. 47 is a picture of a completed double eye bolt.

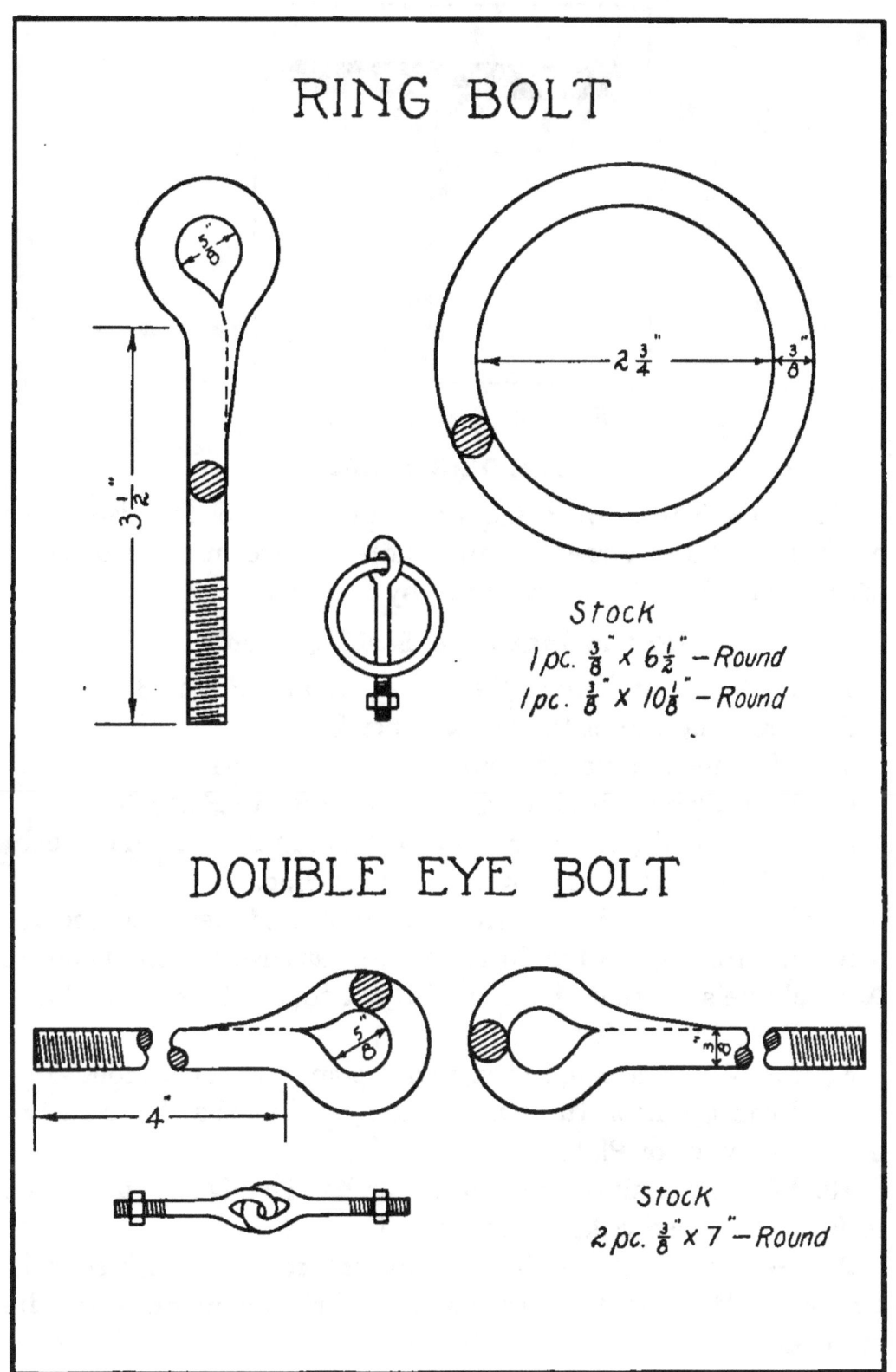

RING BOLT

Stock
1 pc. $\frac{3}{8}$" × 6$\frac{1}{2}$" – Round
1 pc. $\frac{3}{8}$" × 10$\frac{1}{8}$" – Round

DOUBLE EYE BOLT

Stock
2 pc. $\frac{3}{8}$" × 7" – Round

PLATE 23

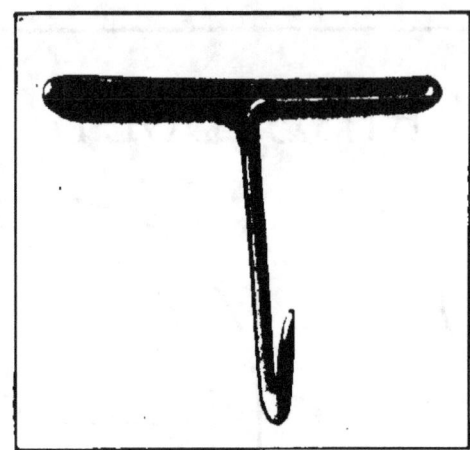

FIG. 48. SCALDING HOOK.

SCALDING HOOK

This hook is used when scalding a pig. One of the size given on Plate 24 has a large enough handle for two men. However, it can be used just as successfully by one man.

How to Make the Scalding Hook

1   First read and study the Lap Weld and Plate **18.**

2.   Scarf one end of the bar and bend at D.

3.   Measure 4⅜ inches from D to C, and mark C.

4.   Bend directly backward at C as shown in Plate 24.

5.   Measure 9½ inches from C to B and mark. Bend directly backward again at B as shown in the drawing.

6.   Measure 4½ inches from B to A and bend as shown. You can probably do this in the vise most easily. The bend at A should be such that the iron will just touch the scarf at D.

7.   Weld the scarfed end

8.   Draw a square and then round point two inches long

9.   Bend the hook three inches long and to the shape shown in the side view of Plate 24.

10. Weld one-half of the double bar handle slightly, **as shown** in Fig. 48.   Then weld the other half.

Note.— Doubling the handle makes it much stiffer and stronger. Welding the two parts slightly again adds to the stiffness.

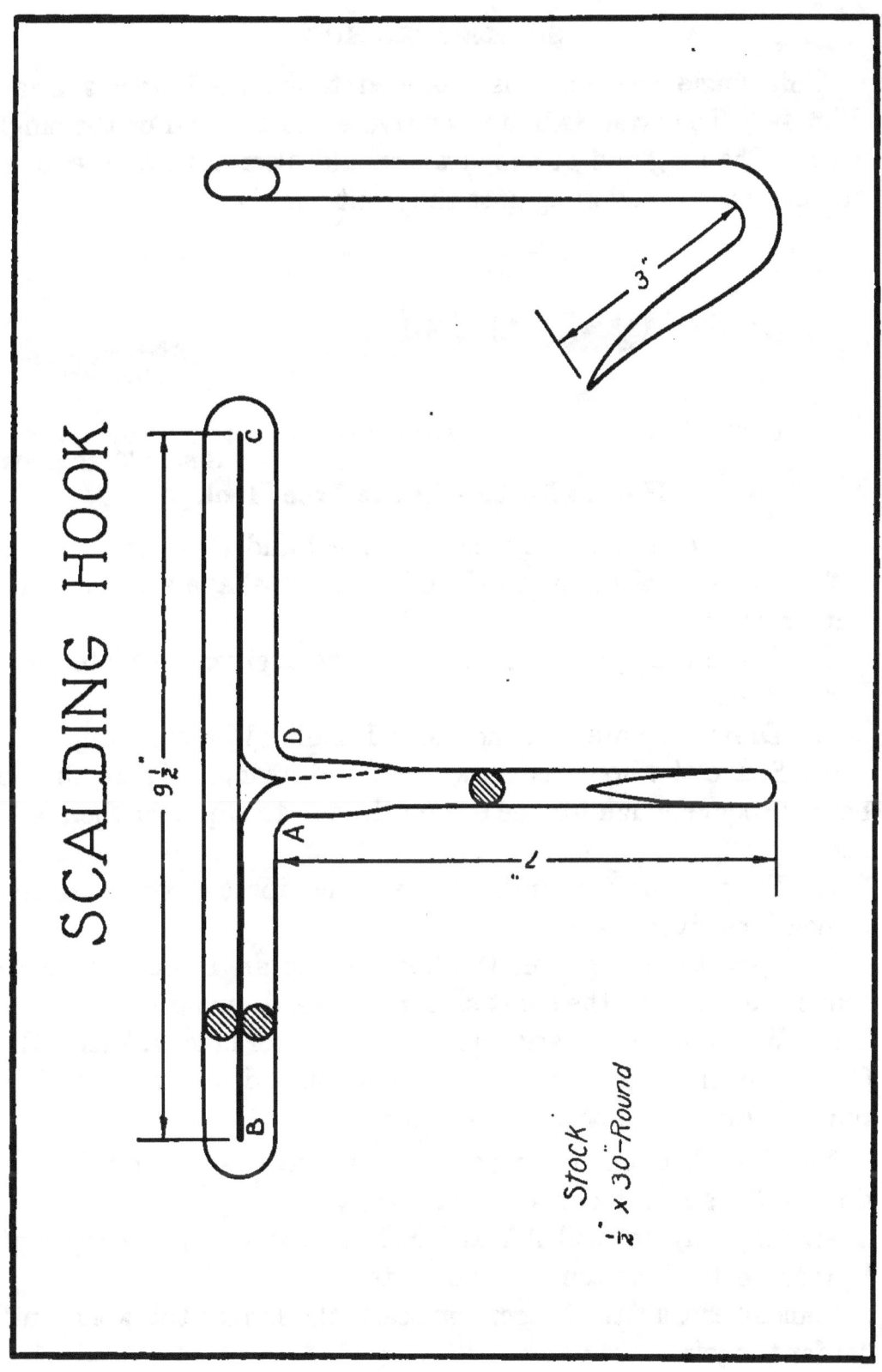

SCALDING HOOK

9½"

3"

7"

Stock
½" x 30"–Round

PLATE 24

SINGLE TREE HOOK

This single tree hook is intended to be used with a strap, Fig. 49. The strap is bolted or riveted to the end of the single tree. The long end of the strap should always be long enough to act as a protection against the front wheels.

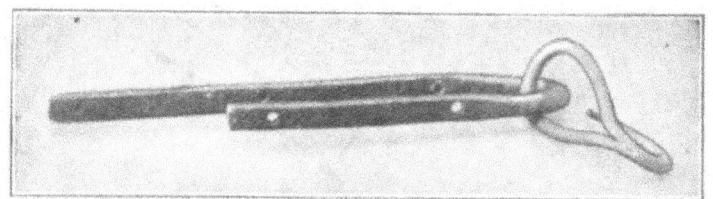 

FIG. 49.   SINGLE TREE HOOK AND STRAP.   FIG. 50.
SINGLE TREE HOOK.

How to Make a Single Tree Hook

1.   First read and study the Lap Weld and Plate 18.
2.   On one end of the stock for the hook shape a rather long narrow scarf.
3.   Shape the eye part to the dimensions given on Plate 25.
4.   Weld.
5.   Draw a square and then round point 1½ inches long.
6.   Shape the hook as shown in the drawing. There should be just room enough for the eye on the tug to slip over the hook. See Fig. 50.
7.   Flatten the two ends of the stock for the strap to the dimensions given.
8.   Bend the strap over the horn of the anvil, being sure to shape the curve on the short side, as in the drawing.
9.   With a square, mark across both ends for holes AA and BB. Center-punch these and the other hole, and drill. AA and BB must be directly opposite each other.

Notes.—When making repairs be sure to change any dimensions of the strap that may be necessary.

An easy way to drill AA and BB is to first slip a couple of ¾-inch boards between the two parts.

Remember that the longer the scarf, the longer the weld, and harder the job.

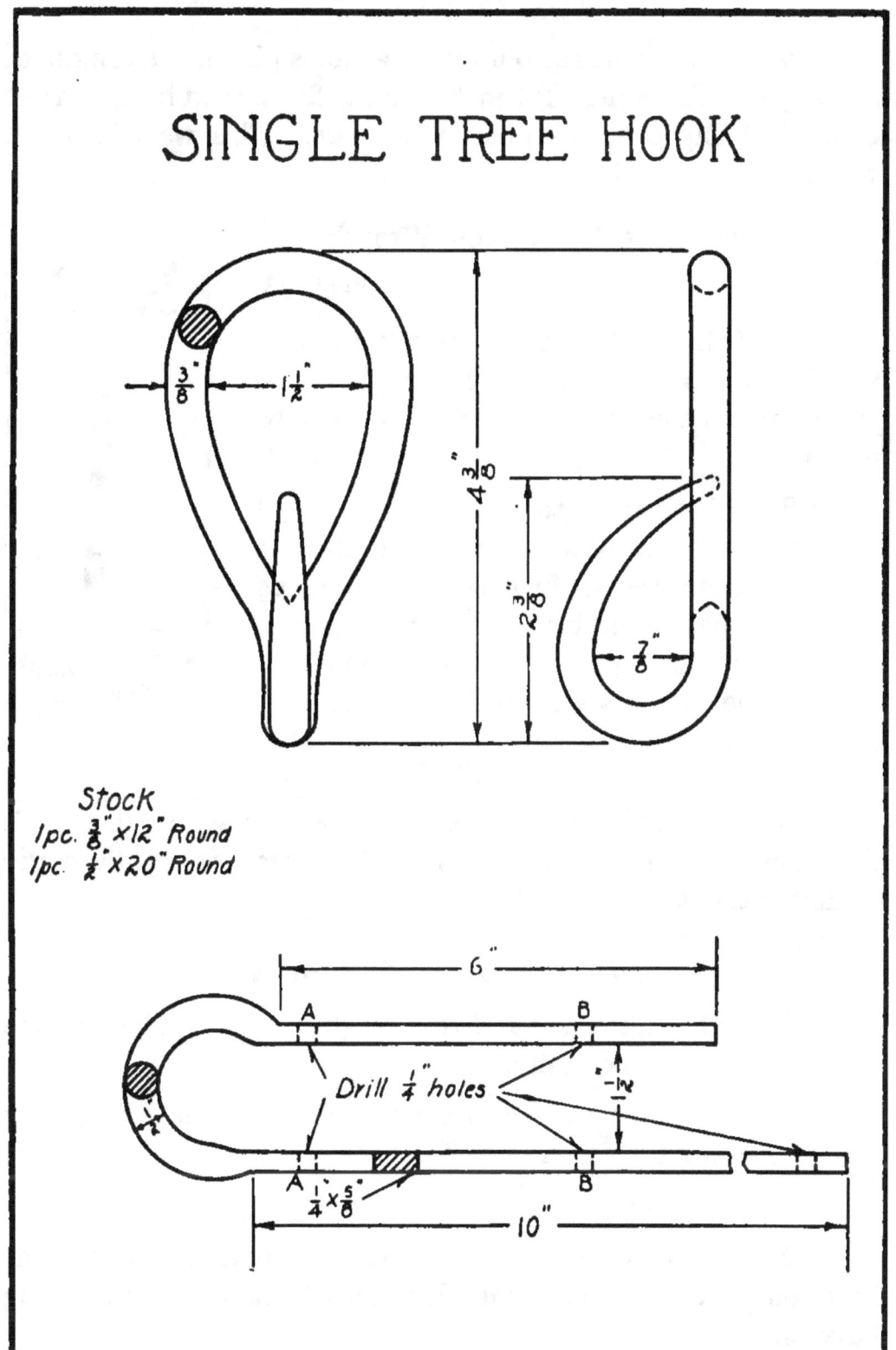

SINGLE TREE HOOK

Stock
1 pc. $\frac{3}{8}$" x 12" Round
1 pc. $\frac{1}{2}$" x 20" Round

Drill $\frac{1}{4}$" holes

$\frac{1}{4}$" x $\frac{5}{8}$"

PLATE 25

WAGON WRENCH

The wagon wrench is also used for a clevis pin on the tongue of the wagon.   The width, 3¼ inches, Plate 26, should be such that it will just slip over the nut on the axle.   This weld is not a true lap weld as neither part is scarfed.

How to Make the Wagon Wrench

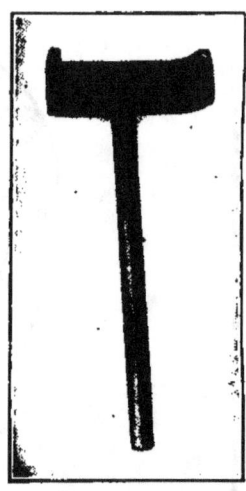

1.   Form the head of the wrench, Plate 26.

2.   Heat the head, and over a cold piece of round iron or with a top fuller form a small groove across the back where the pin will be welded.

3.   Bring both parts to a welding heat (at the same time) and weld.   By placing the pin underneath the groove of the head and picking up the two parts together when removing from the fire, the welding may be made somewhat easier.   The wagon wrench is shown in Fig. 51.

Fig. 51. Wagon Wrench.

BOLT HEAD

If you have made the various clevis pins shown in this book the making of a bolt head is a simple matter.   It can be made readily without a heading tool.

How to Make a Bolt Head

1.   Heat the end of the rod to a bright red heat, place in the vise and upset considerably.

2.   Heat again and place in the vise at the same point, only turned one-quarter around.

3.   Shape the head.

4.   Make a good square shoulder all the way around under the head, on the edge of the anvil   Use a set hammer for this if you have one.

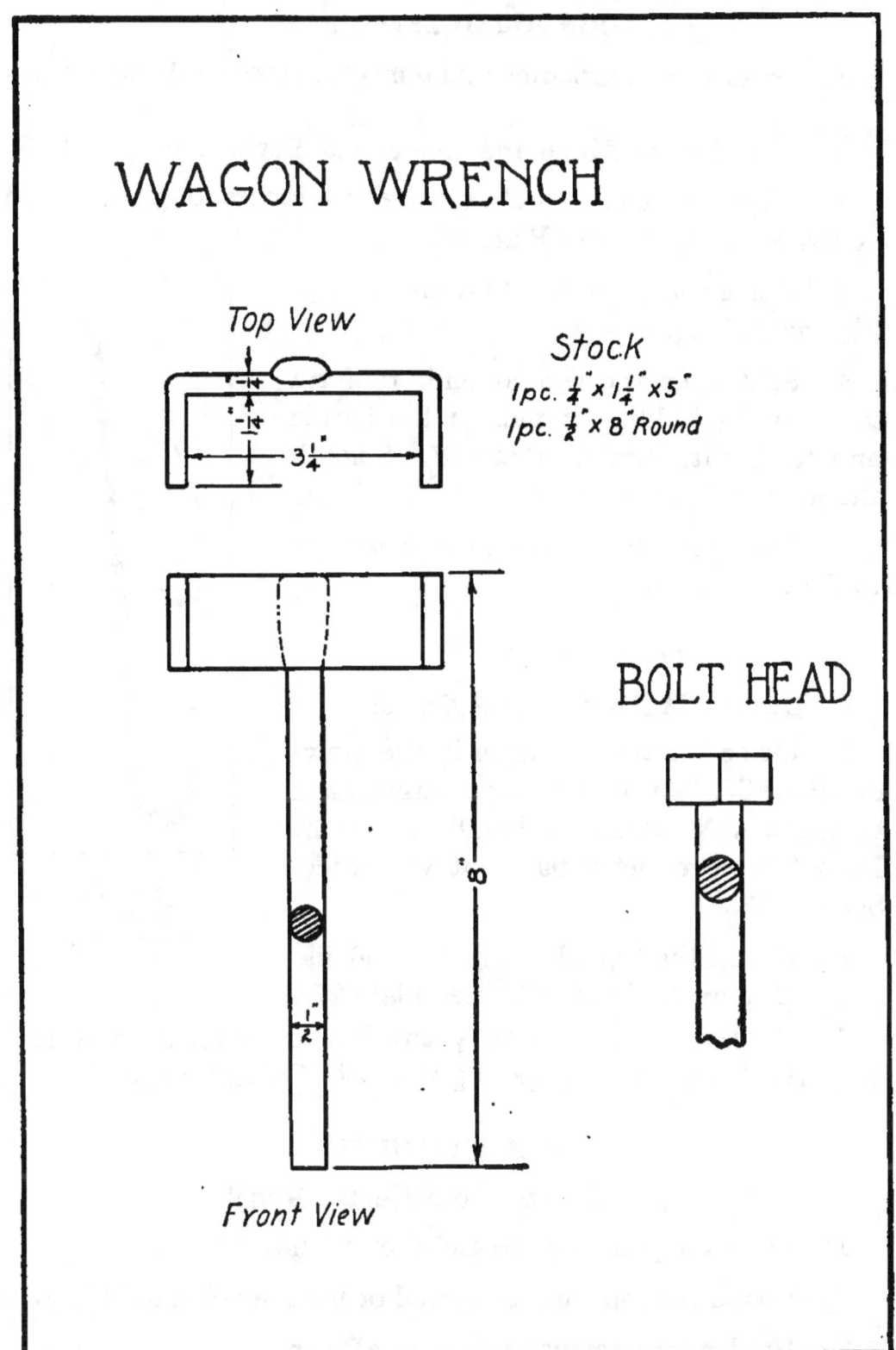

WAGON WRENCH

Top View

Stock
1pc. ¼" x 1¼" x 5"
1pc. ½" x 8" Round

BOLT HEAD

Front View

PLATE 26

COMPOUND LEVER BAR

First read about hardening and tempering tool steel on page 87.

How to Make the Compound Lever Bar

1. Shape one end to a chisel edge two inches long, and bend as shown in Fig. 52 and Plate 27.

2. Shape the other end to a chisel edge 2½ inches long.

3. Split open the second end, to form the claw, by holding it flat on the hardie and cutting it. See A, Plate 27. Smooth the jaws with a file.

4. Bend both ends; harden and temper to dark blue color.

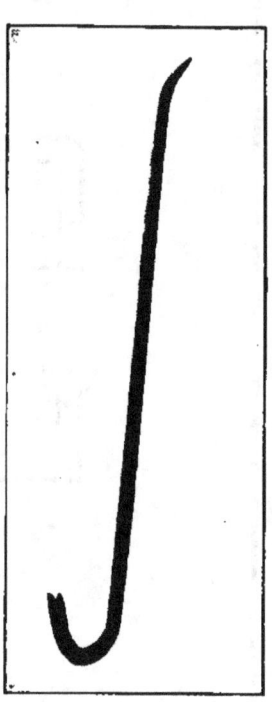

Fig. 52.
COMPOUND
LEVER BAR.

COLD CHISEL

How to Make the Cold Chisel

1. Shape the end as shown in the drawing, Plate 27. The cutting edge must always be made wider than the handle above it. Do not, however, let it flare out very wide. See Fig. 53.

2. Round the top edge slightly and file or grind to an angle of 70°. See Plate 27.

3. Harden, and temper to a dark blue color for general use. See "Hardening and Tempering the Cold Chisel," page 87.

CENTER PUNCH

How to Make the Center Punch

1. Draw the point square and then round.

2. Round the top edge and grind or file the point to 90°.

3. Harden, and temper to a purple color.

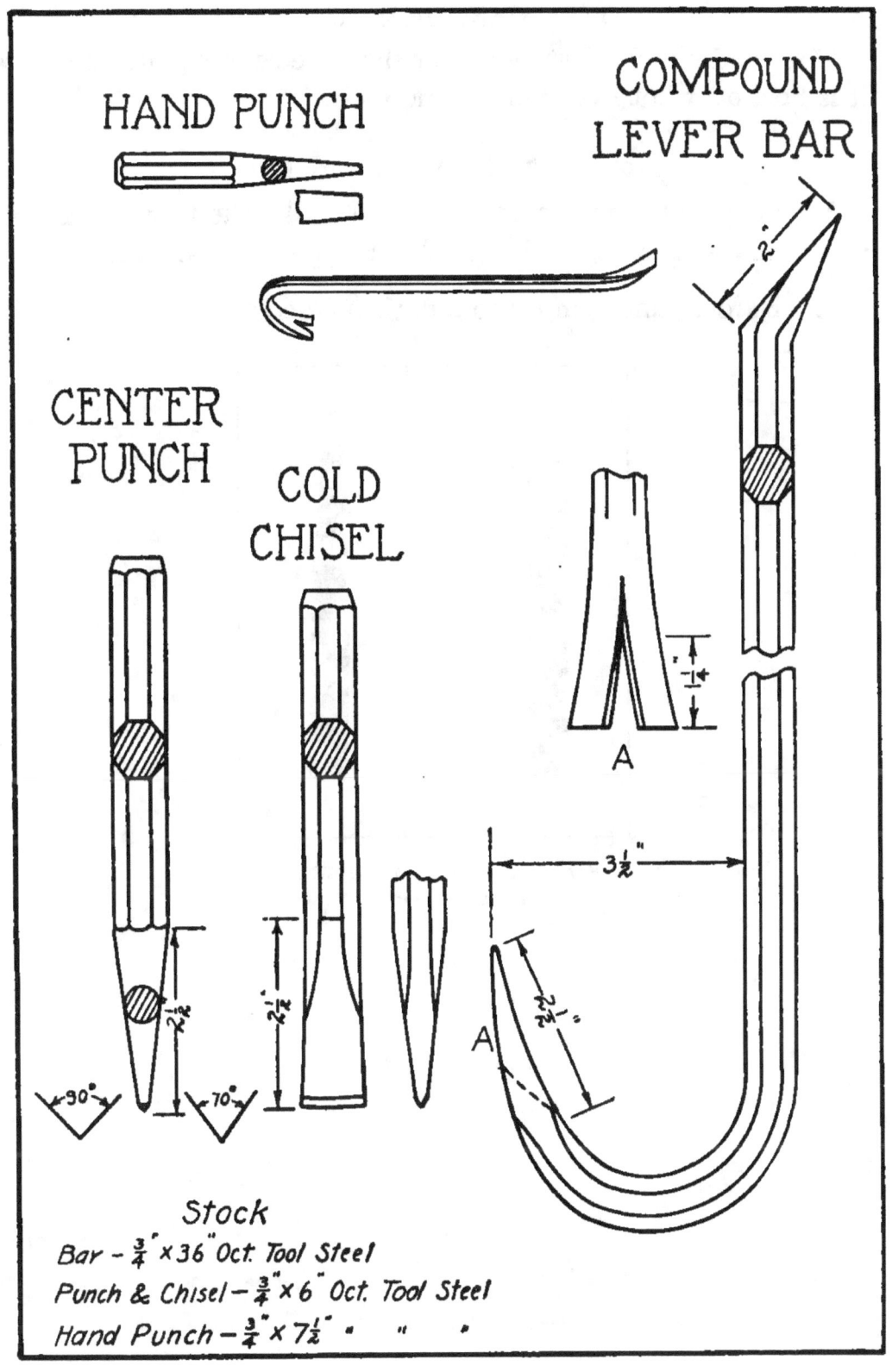

HAND PUNCH

COMPOUND
LEVER BAR

CENTER
PUNCH

COLD
CHISEL

A

3½"

A

90°   70°

Stock
Bar – ¾" × 36" Oct. Tool Steel
Punch & Chisel – ¾" × 6" Oct. Tool Steel
Hand Punch – ¾" × 7½" "   "   "

PLATE 27

HAND PUNCH

The hand punch must be longer than the center punch because it is held over hot iron while being used.

How to Make the Hand Punch

1.　Shape like the center punch but make the taper longer as shown in the drawing and Fig. 53.　The point must be flat.

2.　Harden, and temper to a dark blue color.

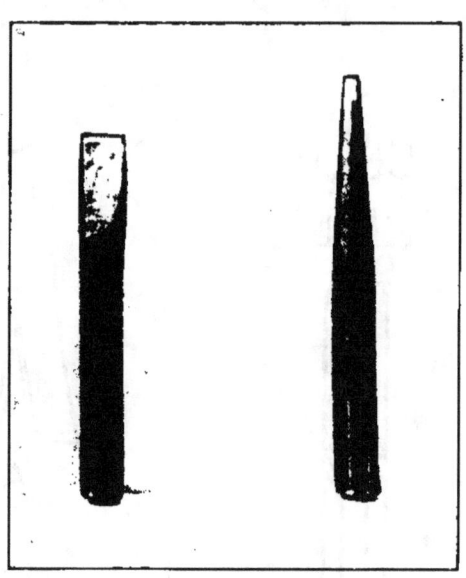

FIG. 53.　COLD CHISEL
AND HAND PUNCH.

ICE TONGS

A good way to determine the curve of the tongs is to mark it on a board or piece of paper. Both parts are made exactly alike, Fig. 54.

How to Make the Ice Tongs

1. Upset that part of the rod, three inches long, Plate 28, to ¾ inch diameter.

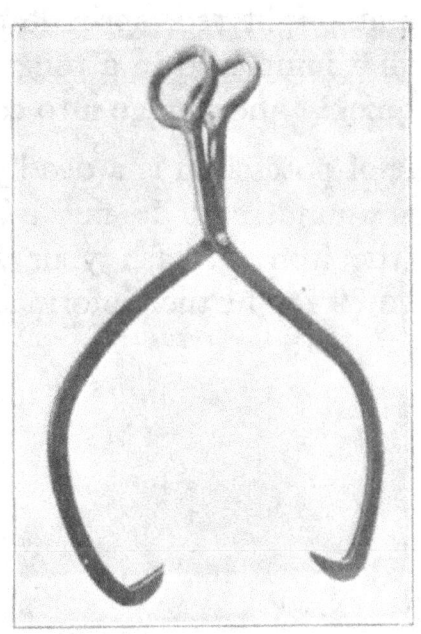

FIG. 54. ICE TONGS.

2. Draw 10 inches of the short end (see drawing on the left of Plate 28) to ½ inch round. This now becomes about 12 inches long.

3. Scarf the end, shape the handle and weld, just as for the baled hay hook, Plate 21.

4. Flatten all of the rest of the rod to ⅜ inch thick. It will be ¾ inch wide except that part which was upset, which will be 1⅛ inches wide. This part will have the hole drilled in it.

5. Draw a square point as shown in the drawing.

6. Bend the shape indicated in Plate 28.

7. Drill the hole, and make a rivet if you have none.

8. Make the second half of the tongs and rivet the two together.

9. Offset the handles a trifle just below the weld so that they will be opposite each other when the tongs are closed.

10. Case harden the points. They cannot be hardened and tempered because wrought iron or mild steel has so little carbon in it. Heat a point bright red, sprinkle cyanide of potassium over it or rub with a lump held in a tongs. When the iron has lost color reheat quickly and plunge into cold water.

Note.— Cyanide of potassium is a deadly poison. Do not let it touch an open wound, and do not inhale the fumes when sprinkling it on the iron. Wash your hands when you are finished. See page 89 for further information about case hardening.

as possible, and yet not so tight that it cannot be freely rotated. This kind of a fit is called a close running fit.

Cutting Threads.—The portion to be threaded should be turned a little smaller than the diameter at the bottom of the threads in piece A. This size is measured by means of the inside spring-thread calipers.

There is no recess, or groove, cut at the end of this thread, so that if the threading tool is allowed to travel farther than the end of the preceding cut, either the point of the tool or the threads may break. To prevent this, the lathe is stopped when the tool is within a half a thread of the end and the cut finished by turning the lathe by hand. In this way the lathe is kept under control and the tool may be drawn back when it reaches the end of the preceding cut. Experienced lathe operators do not, as a rule, turn the lathe by hand, but control the lathe entirely by the shipper.

The tool used for cutting the thread should have the point at one side of the center as shown in Fig. 71, page 58; the reason for this is so that it will cut the thread close to the shoulder.

The speed of the lathe for cutting this thread will be about right for beginners if the belt is on the second smallest step of the cone and the back gears are thrown in.

Finishing the Angle, or Taper.—The 30° angle may be cut by setting the compound rest to the correct angle and using a regular turning tool. In case the tool leaves a few tool marks they may be removed by filing.

If the lathe is not provided with a compound rest, this angle may be cut by setting a square-nose tool, as in Fig. 90, with the aid of a thread gauge. Any other angle would have to be set with a bevel and bevel protractor.

This tool is not as wide as the surface to be cut because one that will cut the full width is very liable to chatter. It is therefore better to make several cuts with a narrow tool. The surface can then be finished smooth by filing.

STRAIGHT LIP TONGS

The straight lip tongs is not difficult to make, once the shapes of the two parts for the jaws are determined. Both are made exactly alike.

FIG. 55. TWO HEADS OF STRAIGHT LIP TONGS.

How to Make the Straight Lip Tongs

1. From a ¾x¾x5-inch piece of iron forge to the shape and dimensions shown in Plate 29 and Figs. 55 and 56. The top and front views of the head, Plate 29, especially should be studied. If necessary, upset the end a trifle to get a full 1-inch width for the lips.

2. File the corners a trifle if necessary, Fig. 55, so that the two parts will fit closely.

3. Round the ends to which the handles are to be welded, and scarf them for lap welds.

4. Scarf one end of each handle, and weld.

5. Drill one hole as shown in the drawing. Place the two jaws of the tongs together in proper position. Mark for the hole in the other jaw. Drill and rivet the two jaws together.

Note.—When the jaws are closed, the lips should be about ¼ inch apart and parallel with each other. This is a good size for general use, but the lips may be reforged at any time for use on thicker or thinner stock. This tongs can also be shaped into a link tongs, Fig. 3.

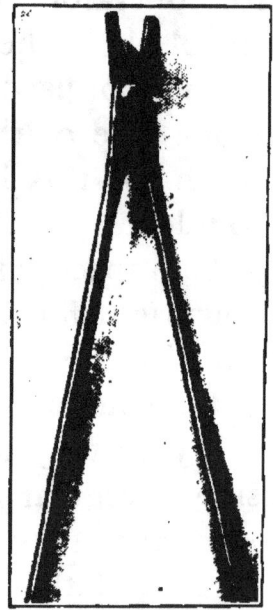

FIG. 56.
STRAIGHT LIP
TONGS.

STRAIGHT LIP TONGS

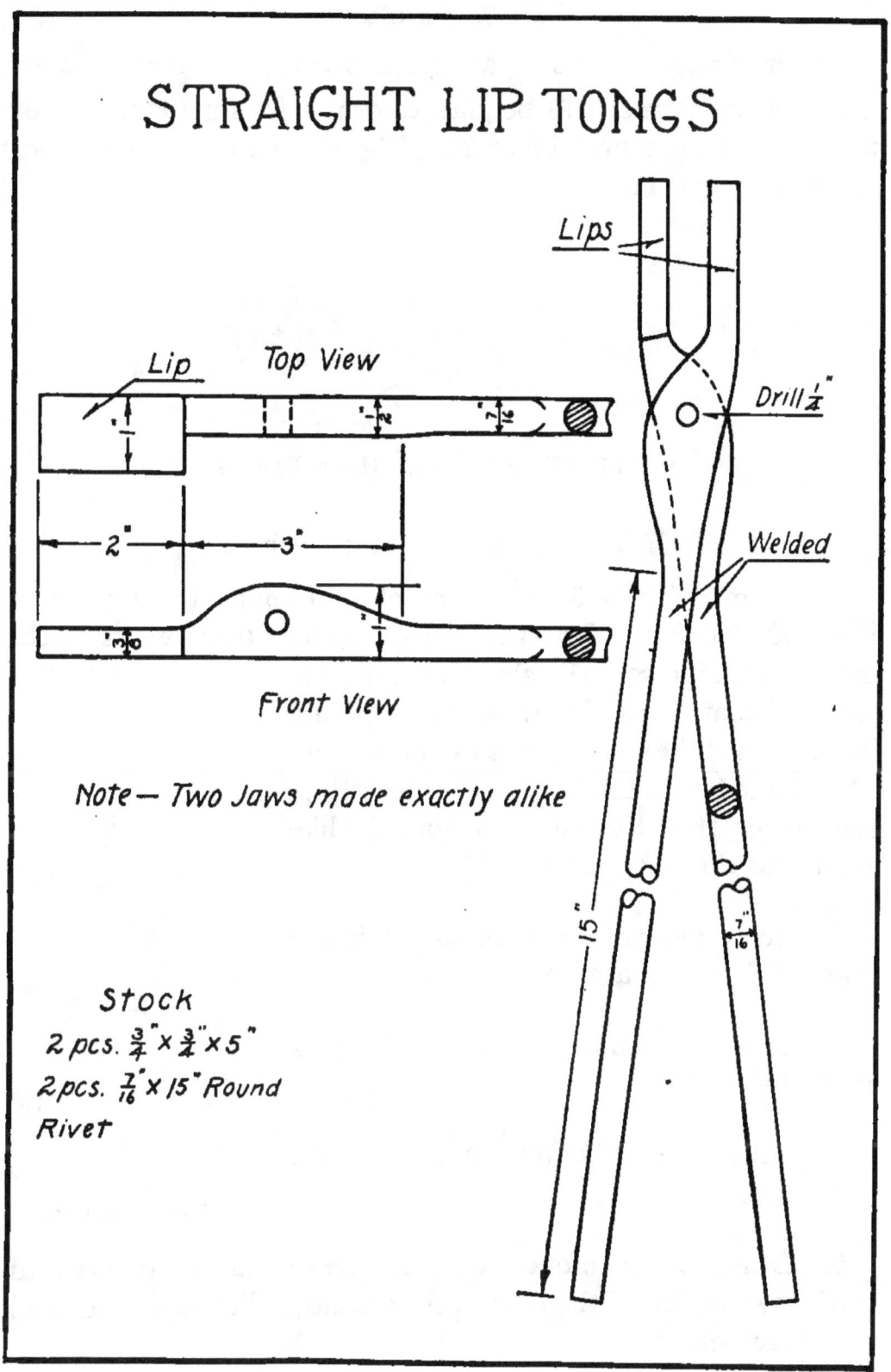

Lips

Drill ¼"

Lip

Top View

Welded

2"   3"

Front View

Note — Two Jaws made exactly alike

15"

7/16

Stock
2 pcs. ¾" X ¾" X 5"
2 pcs. 7/16" X 15" Round
Rivet

PLATE 29

HOOF PARERS

The hoof parers, or hoof paring tongs as it is sometimes called, is a tool any farmer will be glad to own. It is made much like the straight lip tongs, Plate 29. The end of one jaw is sharp and the other is blunt.

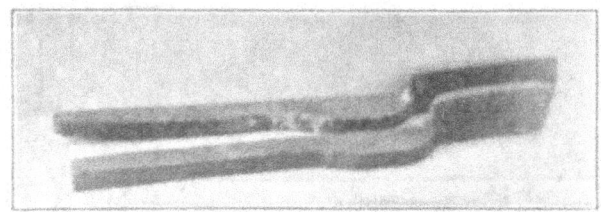

FIG. 57. TWO JAWS OF HOOF PARERS.

How to Make the Hoof Parers

1.   From a ¾x¾x5-inch piece of iron shape the two jaws, Plate 30, and Figs. 57 and 58. The top and front views of the blunt jaw A indicate the shape and dimensions of that jaw. Likewise the top and front views of the sharpened jaw B are shown. The blunted end of A is made by upsetting. Remember that the jaws are forged alike except for the ends; see Fig. 57.

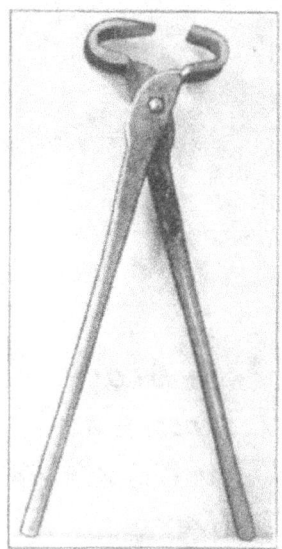

FIG. 58.
HOOF PARERS.

2.   Bend the jaws to the approximate shape shown in Plate 30.

3.   Round the ends to which the handles are to be welded.

4.   Scarf these ends and the ends of the handles, and weld.

5.   Drill the hole in one jaw; place the two jaws together and mark the location of the hole in the second; drill the second, and rivet together.

HOOF PARERS

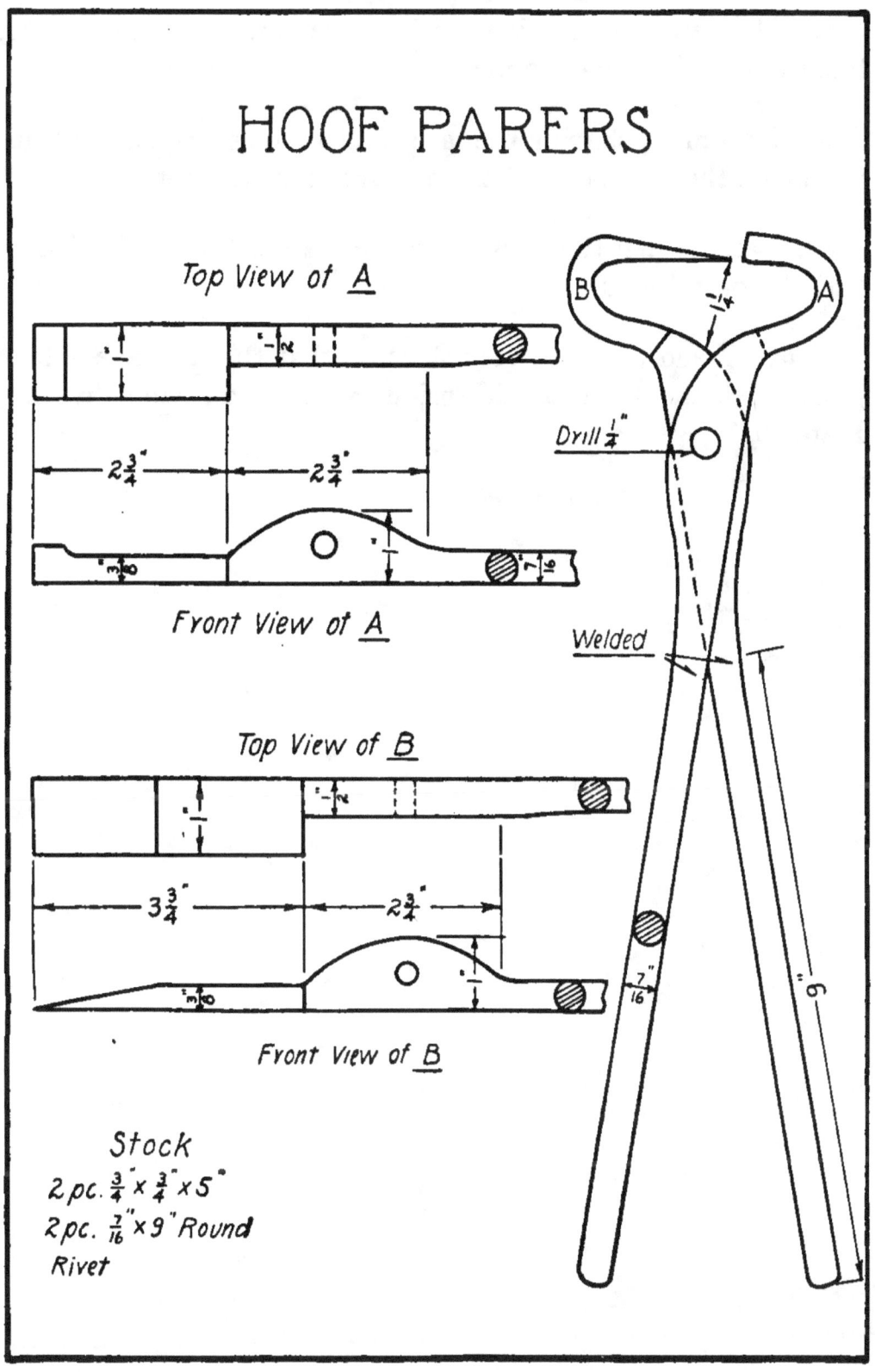

Top View of <u>A</u>

Front View of <u>A</u>

$2\frac{3}{4}$" $2\frac{3}{4}$"

$\frac{1}{2}$"

$\frac{3}{8}$"

$\frac{7}{16}$"

Top View of <u>B</u>

Front View of <u>B</u>

$3\frac{3}{4}$" $2\frac{3}{4}$"

$\frac{1}{2}$"

$\frac{3}{8}$"

B A

Drill $\frac{1}{4}$"

Welded

$\frac{7}{16}$"

9"

Stock
2 pc. $\frac{3}{4}$" × $\frac{3}{4}$" × 5"
2 pc. $\frac{7}{16}$" × 9" Round
Rivet

PLATE 30

6. Sharpen the one jaw with a file, and also square the blunt one in the same manner.

7. Now finish shaping the jaws so that the sharp one will just push past the blunt one when they are pressed together.

8. Case harden the jaws. See the "Notes" under Ice Tongs and also on page 89.

Note.— Bend the jaws very short so that the leverage will be great. See the 1¼-inch dimension on the right-hand drawing, Plate 30.

IRON AND STEEL

All commercial iron and steel contains carbon in varying amounts. Such characteristics as hardness, toughness and fusibility are determined by the amount of carbon contained in the iron. The kinds of iron and steel which the blacksmith most frequently comes in contact with are as follows:

1. Mild steel—contains 1/10 per cent to 1/5 per cent of carbon

2. Wrought iron—contains 1/25 of 1 per cent of carbon.

3. Tool steel—contains 3/5 of 1 per cent to 2 per cent of carbon.

4. Cast iron—contains 2 per cent to 6 per cent of carbon.

The amount of carbon contained in the various kinds of iron and steel is used to help designate it. A "point" is 1/100 of 1 per cent. Therefore, what is known as 20-point mild steel contains 1/5 of 1 per cent carbon. Again, a 150-point tool steel contains 1½ per cent carbon.

Mild Steel—Nearly all iron used today by blacksmiths is known as mild steel or blacksmith iron. It has almost supplanted wrought iron. Mild steel contains considerably more carbon than wrought iron, but not enough to allow for hardening and tempering. Some of it will harden a trifle when heated and plunged into water, but it is not satisfactory for tools. Mild steel can be welded successfully, but not as readily as wrought iron. It is made by the Bessemer process which is the quickest and cheapest process used in iron making. Mild steel is made in rods, bars, bands and sheets and is sold by the pound. The shapes are known as rounds, squares and flats.

Wrought Iron—Wrought iron is light gray or silver in color, and is fibrous or stringy. It can be bent, twisted and hammered cold more successfully than any other iron. After cold manipulation it can be returned to its original condition by annealing it. This is done by heating to a good red heat and allowing it to cool slowly without water.

Wrought iron is easily forged when heated, and is the best kind of iron to use for welding. A very high grade known as "Norway" or "Swedish" iron is the best that can be purchased. It is higher in price, but of late is almost unobtainable. Mild steel or blacksmith iron has now practically taken the place of wrought iron, for all smithing purposes. Wrought iron is manufactured in round, square and flat rods.

Tool Steel—Tool steel has enough carbon in it so that it can be hardened and tempered. In addition to having the correct shape a tool must have the proper degree of hardness for the use for which it is intended. Several different tools all tempered to a different degree of hardness may be made from the same bar of tool steel. The principal shapes in which tool steel is made are square, rectangular (flat) and octagonal bars. It is sold by the pound.

When tools forged from tool steel crack, bend or are blunted, the steel should not necessarily be considered defective. The trouble can probably be traced to the workman who forged, hardened, and tempered the tools.

Tool steel cannot be heated as highly for forging as wrought iron or mild steel It "burns" at a glowing white heat and crumbles. It should be heated slowly, and frequently if necessary, to a good red heat—no higher—and forged. Use little draft and turn frequently. Do not upset tool steel any more than is absolutely necessary as cracks are likely to follow. A good tool steel is bluish gray in color. When possible, select 70 to 90-point steel for making blacksmith tools. Also get it in the size and shape which is nearest the finished tool.

Cast Iron—Cast iron cannot be forged. When heated to a bright red or yellow heat it crumbles. It is used for many "parts" on farm implements. Cast iron can be readily drilled and filed. It is usually gray in color, and cracks easily under a blow. A blacksmith does not come in contact with it as frequently as other kinds of iron.

HARDENING AND TEMPERING

Hardening is making the tool steel as hard as possible, and tempering is reducing the hardness to a certain definite desired degree.

Hardening—Heat the tool steel to a blood red heat and plunge into clean, cool water. The addition of salt in the water has a tendency to make the hardness greater. The water should be kept cool by changing or adding, when numerous articles are hardened. The steel is now as hard as it can be made, and it cannot be filed.

Tempering—After a piece of tool steel has been hardened, a certain definite degree of hardness may be obtained by reheating it to a given temperature. The temperature is determined by the color on a polished surface of the tool. The use for which a tool is intended determines the degree to which it is to be tempered, and therefore the color to which it is to be heated.

The first color to appear on the polished surface, when reheating for tempering, is light straw. At this color the steel is tempered just a trifle. After light straw the colors appear in the following order: dark straw, brown, purple, dark blue and light blue. A light blue color indicates that the steel has been tempered to the softest point before·it returns to its original condition.

Many tools can be hardened and tempered in one heating. The cold chisel, Plate 27, is such a tool.

Hardening and Tempering the Cold Chisel—First heat about 2 inches or $2\frac{1}{2}$ inches of the cutting end to a blood red heat, using a slow fire with little draft. Plunge the point into cool water to a depth of about 1 inch. Move backward and forward, and up and down slightly. This movement prevents the steel from hardening to a definite water line. Such hardening to a water line usually results in a break when the chisel is struck. When the 1 inch or so of the point has been thoroly cooled, remove from the water and quickly polish with a piece of emery cloth, tacked on a stick. The heat remaining in the upper part of the chisel will now begin to run down into the cooled and hardened end, and the tempering will have begun. The colors will now appear. The first color to be noted on the cutting edge will be light straw. Above this will be dark straw, brown, purple, dark blue and light blue. The colors move down as the heat from above runs down. The light straw will disappear on the edge and the dark straw above it will take its place. Then the dark straw will disappear, and so on until the dark blue color reaches the cutting edge. The moment the edge is dark blue plunge into cool water.

For general use a dark blue is the proper color to temper a cold chisel. A hard test is to hammer it straight into a piece of iron. If the edge does not bend or crack the tool will give good service. The center punch, Plate 27, is hardened and tempered in the same manner.

Tools and Their Tempering Colors—The farm blacksmith has little call for tempering to any of the harder colors. Light and dark straw and brown are used principally on machinist's tools and fine instruments.

Brown—Used for ball pein and other heavy hammers.

Purple—Used on center punches, stone drills, occasionally on cold chisels, and the faces of riveting hammers.

Dark blue—Used on cold chisels, lever or wrecking bars, some knives, blacksmith's cold chisels or cutters, hardies when used for cutting cold stock, and the pein of a riveting hammer.

Light blue—Used on screwdrivers.

Tools used for cutting or shaping hot stock, such as punches, hot chisels and hardies, do not require hardening and tempering.

Annealing—Sometimes it is desirable or necessary to bring the steel back to its original condition. This is a softening process and is sometimes used after a tool has been forged, and before it is hardened and tempered. One way to anneal is to heat the iron to a good red heat and bury it in dry lime or ashes and allow it to cool. A decidedly quicker method is to heat to a good red color and allow it to cool in the air until no heat color is visible when held in a dark corner of room. When this point is reached cool entirely by plunging into water. Annealing is done to eliminate any strains in the steel such as might be caused by hammering, bending or hardening.

Case Hardening—Case hardening is a process of hardening the surface of wrought iron or mild steel, which cannot be hardened and tempered like tool steel. Where no special furnace and equipment is available the method described under Ice Tongs may be used. Case hardening simply hardens a thin layer of the outside surface. By repeating the process several times the thickness of the hardened part may be materially increased.

Pay especial attention to the "Notes," Page 78, cautioning against poisoning from the cyanide of potassium and its fumes.

It is a good plan for the farmer to save such bolts, nuts, rods, sheet iron, etc., which may come to hand in one way of another. Such things can sometimes be used as they are, or can easily be changed to a size or shape needed. A small supply of various sizes of screws, nails, bolts of various kinds, pipe, rivets and iron rods should be kept in the farm shop so that repairs and new work can proceed without the farmer first having to go to town for materials.